Soft Computing Applications for Advancements in Power Systems

RIVER PUBLISHERS SERIES IN POWER

The "River Publishers Series in Power" is a series of comprehensive academic and professional books focussing on the theory and applications behind power generation and distribution. The series features content on energy engineering, systems and development of electrical power, looking specifically at current technology and applications.

The series serves to be a reference for academics, researchers, managers, engineers, and other professionals in related matters with power generation and distribution.

Topics covered in the series include, but are not limited to:

- Power generation;
- Energy services;
- Electrical power systems;
- Photovoltaics;
- Power distribution systems;
- Energy distribution engineering;
- Smart grid;
- Transmission line development.

For a list of other books in this series, visit www.riverpublishers.com

Soft Computing Applications for Advancements in Power Systems

Editors

Vijay Kumar Sood

Ontario Tech University, Canada

Krishna Murari

The University of Toledo, Ohio, USA

Om Hari Gupta

National Institute of Technology Jamshedpur, India

Anupam Kumar

National Institute of Technology, Patna, Ashok Rajpath,
Bihar, PIN - 800005, India

Published 2024 by River Publishers
River Publishers
Alsbjergvej 10, 9260 Gistrup, Denmark
www.riverpublishers.com

Distributed exclusively by Routledge
605 Third Avenue, New York, NY 10017, USA
4 Park Square, Milton Park, Abingdon, Oxon OX14 4RN

Soft Computing Applications for Advancements in Power Systems / by Vijay Kumar Sood, Krishna Murari, Om Hari Gupta, Anupam Kumar.

Routledge is an imprint of the Taylor & Francis Group, an informa business

ISBN 978-87-7004-141-6 (hardback)
ISBN 978-87-7004-701-2 (paperback)
ISBN 978-87-7004-687-9 (online)
ISBN 978-87-7004-686-2 (master ebook)

Contents

Preface

This book covers the applications of soft computing techniques for advancements in power systems. Power systems have grown from a simple radial system with a few generators and dedicated loads to massive, complex, and integrated power systems serving the needs of modern societies. The rapid growth in power systems took place after World War II ended in 1945, as AC lines were required from remote generating sites to load centers. The AC long-distance lines required ever increasing higher voltages to reduce the transmission power losses. The early 1950s also saw the introduction of High Voltage DC transmission systems in Sweden using mercury-arc converters for the special needs of bulk power transmission over long distances and undersea interconnections.

In 1958, General Electric introduced the Silicon Controlled Rectifier (SCR). By the late 1960s, the introduction of power electronics to power systems brought about the era of static var compensation and flexible AC transmission systems (FACTS). In the early 1960's, the development of the Electromagnetics Transients Program (EMTP) program by H. Dommel also made it easier to simulate and understand complex power systems. In the early 1970s, the load flow programs were being developed at Bonneville Power Administration (BPA). In 1984, the personal computer (PC) was introduced by IBM. By the late 1980's, the widespread merging of power electronics, information technology and communications facilities brought about a revolutionary change to the industry as the control and protection of the ever-growing power systems became increasingly important and complicated.

Today, converters are widely used in power systems, renewable energy systems, electric vehicles as well as smart home appliances. As the needs have evolved, new multi-level converter topologies have been developed. For these different topologies, advanced pulse width modulation (PWM) techniques and control systems are required to generate firing signals for the power electronic switches. Also, as new converter topologies come with challenges like DC voltage balancing, circulating currents, more available switching redundancies, harmonic generation etc. the need for sophisticated control systems has grown.

Converters can be controlled using either analog, digital or hybrid systems. In the early days, analog controllers were used but were soon replaced by digital controllers in the 1970s. Digital control systems offered advantages, such as elimination of operating point variations due to analog device aging and temperature drift effects, ease of implementing sophisticated algorithms and control laws to improve dynamic performance and efficiency of power converters. They were also easily reconfigured and scaled, compared to analog systems. However, discretized digital control systems bring additional challenges in power converters i.e. resolution limitations, cost and power consumption incurred from the required high-speed, high-resolution analog to digital (A/D) converters and digital PWM techniques. Since there is an inherent delay in the discretized control loop, it introduces errors in output voltage waveform and affects the stability of the entire system. The fully digital controller was built at the end of 1990s with the availability of low-cost microcontrollers and digital signal processors (DSPs).

The era of climate change, environmental concerns, and the gradual replacement of fossil fuels as the source of electrical energy has now reached the point where the smart grid with the use of the internet is now becoming highly necessary to any national grid. The modern power system is now intimately involved with power electronics, HVDC systems, microgrids, power system protection, and energy scheduling to reduce the cost of electricity and provide a stable controllable power supply.

This book will aid researchers who plan to work in the domain. The book will be useful also to industrial practitioners. To the best of our knowledge, none of the existing books related to power systems considers these individual perspectives. Although some of the available books are focused on specific technical subdomains, they lack the wider perspective to meet the requirements of all potential users. The book is unique and one of a kind, as it will have information that is neither easily accessible nor available in a concise format by researchers.

This book is composed of ten carefully selected and diverse chapters by various authors to highlight the impact of soft computing methods in power systems. On behalf of all the co-editors, and authors, it gives me great pleasure to offer this textbook for your usage.

Vijay K. Sood
3 June 2024
Oshawa, Ontario.

List of Figures

List of Tables

List of Contributors

Balasubramanyam, Priyadarshini, *Faculty of Engineering and Applied Science, Ontario Tech University, Canada*

D, Giribabu, *Department of Electrical Engineering, MANIT, India*

Gawre, Suresh Kumar, *Department of Electrical Engineering, MANIT, India*

Gupta, Om Hari, *Department of Electrical Engineering, NIT Jamshedpur, India*

Kamalasadan, S., *Department of Electrical and Computer Engineering, University of North Carolina at Charlotte, United States of America*

Kumar, Anupam, *National Institute of Technology, Patna, Ashok Rajpath, Bihar, PIN - 800005*

Kumar, Jitendra, *NIT Jamshedpur, India*

Kumar, Shailendra, *Department of Electrical Engineering IIT Bhilai, India*

Kumar, Shitanshu, *Department of Electrical Engineering, MANIT, India*

Mahajan, S., *Ontario Tech University, Canada*

Murari, Krishna, *Electrical and Computer Engineering Department, University of Toledo, Ohio, United States of America*

Padhy, N.P., *Department of Electrical Engineering, Indian Institute of Technology Roorkee, India*

Sahu, Sameep, *Department of Electrical Engineering, NIT Jamshedpur, India*

Sharma, G., *Ontario Tech University, Canada*

Sharma, M., *Ontario Tech University, Canada*

Shrivastava, Shashank, *Department of Electrical Engineering, MANIT Bhopal, India*

Singh, Sanjeev, *Department of Electrical Engineering, MANIT, India*

Sood, Vijay Kumar, *Ontario Tech University, Canada*

Suresh, M., *Electrical Engineering Department, Indian Institute of Technology Roorkee, India*

Tiwari, Ravi Shankar, *NIT Jamshedpur, India*

Tripathy, Manoj, *Electrical Engineering Department, Indian Institute of Technology Roorkee, India*

List of Abbreviations

AC	Alternating current
ACO	Ant colony optimization
AGC	Automatic generation and control
AMI	Advanced metering infrastructure
ANFIS	Adaptive neuro-fuzzy inference system
ANN	Artificial neural network
ANSI	American National Standards Institute
BBC	Buck-boost converter
BESS	Battery energy storage system
BFS	Backward-forward sweep
BIBC	Bus-injection to bus-current
BIBV	Bus-injection to bus-voltage
BS	Backward sweep
CC	Constant current at rectifier
CCVT	Capacitor coupling voltage transformer
CEA	Constant extinction angle
CF	Commutation failure
CHIL	Controller hardware-in-loop
CHP	Combined heat and power
CIA	Confidentiality, integrity, and availability
CIA	Constant ignition angle
CM	Common mode
CPN	Counter propagation network
CPS	Cyber–physical system
CSI	Current source inverter
CT	Current transformer
DC	Direct current
DCLG	DC line-to-ground fault
DER	Distributed energy resources
DERMS	Distributed energy resource management system
DG	Distributed generation

DLG	Double line-to-ground
DMR	Dedicated metallic return
DMS	Distribution management system
DoS	Denial-of-service
DRMS	Demand response management system
DS	Distribution system
DSO	Digital storage oscilloscope
EMI	Electromagnetic interference
EMS	Energy management system
ESS	Energy storage system
FACTS	Flexible AC transmission system
FAN	Field area network
FC	Fault classification
FL	Fuzzy logic
FLS	Fuzzy logic system
FS	Forward sweep
GAE	Genetic and evolutionary
GD	Gradient descent
GS	Gauss–Siedel
GTO	Gate turn-off
GUI	Graphical user interface
GWO	Grey wolf optimization
HAN	Home area network
HERIC	Highly efficient and reliable inverter concept
HFPSOMCS	Hybrid feedback PSO-MCS
HIL	Hardware-in-loop
HUT	Hardware under test
HVDC	High voltage direct current
Hy-HVDC	Hybrid HVDC system
I/O	Input/output
IAC	Ignition angle control
IACS	Industrial automation and control system
ICS	Industrial control system
ICT	Information and communication technology
IDS	Intrusion detection system
IEC	International Electrotechnical Commission
IED	Intelligent electronic device
IEEE	Institute of Electrical and Electronics Engineers
IGBT	Insulated gate bipolar transistor

IP	Internet Protocol
ISO	International Organization for Standardization
ISP	Independent system planner
IT	Information technology
ITU	International Telecommunication Union
KCL	Kirchhoff's current law
LAN	Local area network
LC	Load current
LCC	Line commutated converter
LF	Load flow
LL	Line-to-line
LP	Load power
MAN	Metropolitan area network
MEMS	Microgrid energy management system
MF	Membership function
MLF	Multilayer feed-forward
MLP	Multilayer perceptron
MPPT	Maximum power point tracking
MRA	Multi-resolution analysis
MT-HVDC	Multi-terminal HVDC system
MTU	Master control unit
NAN	Neighbourhood area network
NERC	North American Electric Reliability Corporation
NGG	Natural gas-fired generator
NIST	National Institute of Standards and Technology
NN	Neural network
NOC	Normal operating condition
NR	Newton–Raphson
OHL	Overhead lines
OLG	One line-to-ground
OLTC	On-load taps changer
OSI	Open systems interconnection
OT	Operational technology
PHIL	Power hardware-in-loop
PI	Proportional integral
PLL	Phase locked loop
PMU	Phasor measurement unit
PSB	Power swing blocker
PSCAD	Power System Computer Aided Design

PSO	Particle swarm optimization
pu	per-unit
PV	Photo voltaic
PWM	Pulse width modulation
RBF	Radial basis function
RBFNN	Radial basis function neural network
RELU	Rectified linear activation
RES	Renewable energy resources
RMSE	Root mean square error
ROW	Right of way
RPP	RMS pre-processor
RTU	Remote terminal unit
SANS	SysAdmin, audit, network and security
SCADA	Supervisory control and data acquisition
SCR	Short circuit ratio
SDH	Synchronous digital hierarchy
SDO	Standards development organization
SGN	Smart grid network
SIL	Surge impedance loading
SLD	Single-line diagram
SLG	Single line-to-ground fault
SNSC	Superimposed negative sequence current
SNSV	Superimposed negative sequence voltage
SONET	Synchronous optical network
SPSC	Superimposed positive sequence current
SPSV	Superimposed positive sequence voltage
STATCOM	Static VAR compensator
STEPS	Stated policies scenario
SVC	Static var compensator
SVM	Support vector machine
Sw	Switch
TCPS	Thyristor control phase shifter
TCR	Thyristor controlled reactor
TCSC	Thyristor-controlled series compensator
TL	Transformerless
TLIT	Transformerless inverter topology
TSC	Thyristor switched capacitor
UPFC	Unified power flow controller
UPQC	Unified power quality conditioner

VAR	Volt Ampere reactive
VDCOL	Voltage-dependent current-order limiter
VSI	Voltage source converter
WAN	Wide area network
WASA	Wide area situation awareness
WDM	Wavelength division multiplexing

1

Introduction

Krishna Murari[1], Om Hari Gupta[2], Vijay Kumar Sood[3]

[1]Electrical and Computer Engineering Department, University of Toledo, Ohio, United States of America
[2]Department of Electrical Engineering, NIT Jamshedpur, India
[3]Ontario Tech University, Canada
Email: krishna.murari@utoledo.edu, omhari.ee@nitjsr.ac.in, vijay.sood@ontariotechu.ca

Abstract

The importance of soft computing methods has grown over the years, influencing all pertinent fields of science and engineering. Soft computing methods, inspired by the human mind and biological behaviour, have proven to be excellent tools to overcome the difficulties faced in a vast variety of applications in existing power systems. In recent times, power electronics-based converters with distributed generations and loads are being deeply inducted to the existing power system. With the inclusion of these modern power system components, the applications of soft computing in power systems will have to undergo major modifications or upgrades in providing solutions for the modern power system problems. This chapter provides the information regarding the power system areas where power electronics is making a significant impact. Subsequently, an outline provides information regarding the various problems of the modern power system that have been addressed through soft computing methodologies.

Keywords: Power electronics, soft computing, power generation, power transmission and distribution, storage, DC loads, power quality

1.1 Background and Motivation

The importance of soft computing methods has grown over the years, influencing all pertinent fields of science and engineering. One of the relevant fields is power systems where solutions for optimal planning, analysis, operation, and control have been greatly influenced by the application of various soft computing methodologies [1]. Soft computing offers an effective solution for studying and modelling stochastic behaviour, imprecision, uncertainty, decision-making, partial truth, and approximation problems, making them more valuable than traditional techniques. The soft computing methods, inspired by the human mind and biological behaviour, have proven to be excellent tools to overcome the difficulties faced in a vast variety of applications in power systems.

In the recent times, power electronics-based converters with distributed generations and loads are being deeply inducted to the existing network [2].

Power electronics is commonly applied in power systems and promotes expansion towards a more sustainable direction (Figure 1.1). Power electronics technology is still an emerging technology, and it has found its way into many applications, from renewable energy generation (i.e., wind power and solar power) to electrical vehicles (EVs), biomedical devices, and small appliances such as laptop chargers [3]. In the near future, electrical energy will be supplied, handled, and consumed through the usage of power electronics converters. This will not only intensify the role of power electronics technology in the power conversion processes, but also implies that power systems are undergoing a paradigm shift, from centralized distribution to distributed generation [4]. Today, more than 1000 GW of renewables (solar and wind) have been installed, all of which are handled by power electronics technology [5]. More precisely, power electronics have found applications in the following significant areas:

(A) Power generation: Traditionally, electrical power has been primarily generated by burning a fossil fuel and producing steam, which then rotates a turbine generator to generate power [4]. The speed of rotation also decides the frequency of the generated supply. The grid health greatly depends upon the stability of frequency at which power is being generated. For example, if there is an excessive amount of load then energy is taken from the grid at a faster rate than at which energy is being supplied. Therefore, the turbines or movers slow down, and as a result, the AC frequency drops. Since the turbines are large rotating masses with inertia, they oppose changes in the speed (frequency). At

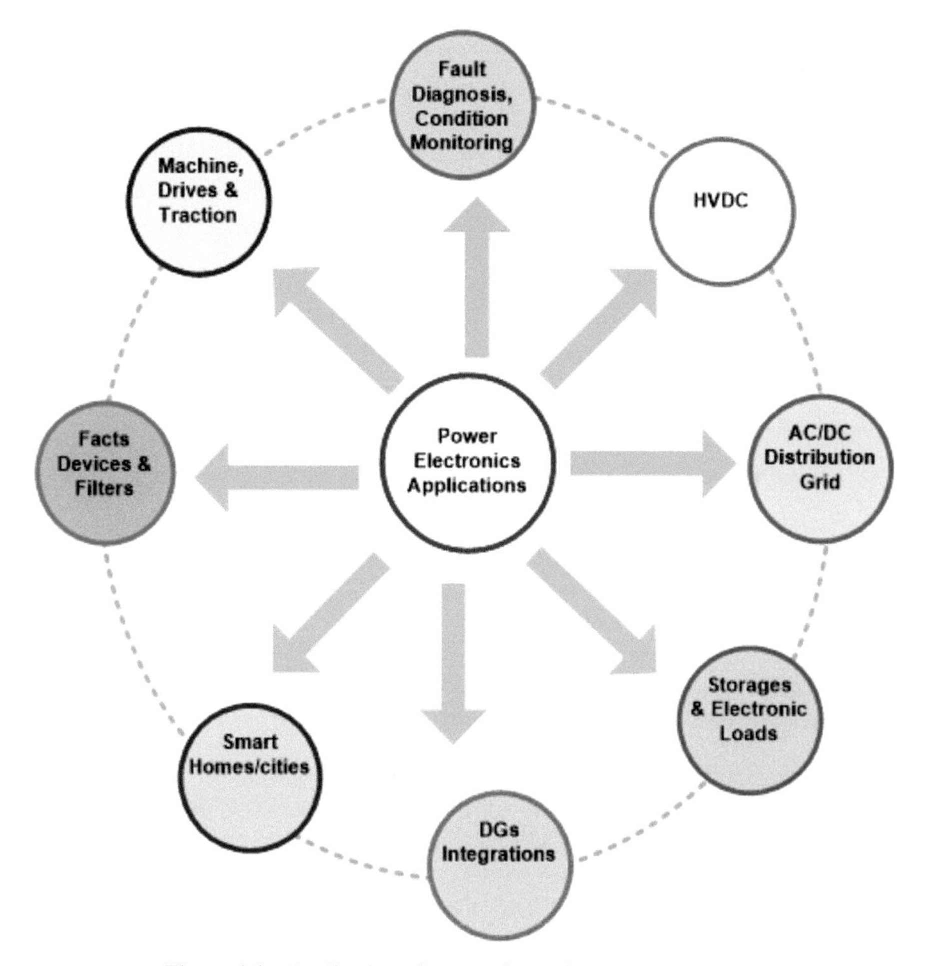

Figure 1.1 Application of power electronics to power systems.

the present time, more inverter-based distributed generations (mostly renewable based) are being integrated to the power grid [6]. The DGs with inverters can generate power at any frequency and they have different inertial properties as compared to steam-based turbine generators. Therefore, to transform the grid with a greater number of inverter-based resources, smarter and more efficient converters are required. The inverter should be capable to respond to any frequency change and to other disturbances that may occur during grid operation [4].

(B) Power transmission and distribution: With the utilization of power electronics-enabled generations and loads, many developments have been made in the areas of transmission and distribution systems. One of the examples, in the transmission system, is high voltage direct current (HVDC) transmission [7], while in the distribution system it is the microgrid [8]. Power electronics was applied to HVDC transmission originally as power system developed in the 1950s. Power energy generated by AC generators is rectified by thyristor converters. Compared to AC transmission, HVDC is more efficient for long-distance and bulk power transmission or submarine or underground cable transmission, power grid connection, and system control [9–11]. The recent advances in low- and medium-voltage fast-switching semiconductors have made it possible to realize advantages of DC distribution over AC distribution [2, 12]. The advantages of DC distribution system are:

- Bidirectional power flow is possible;
- Reactive voltage drop is eliminated in a DC distribution system;
- Synchronization of the multiple loads and sources (phase angles) is not required;
- Higher power transfer capability is achieved as compared to an AC distribution system;
- Greater energy efficiency;
- Higher power quality.

However, complete replacement of AC distribution system with DC is not feasible at this stage since the electric power system is still dominated by AC systems and the lack of economic DC breakers for the protection of the DC grid. As a result, the concept of the AC-DC distribution network (Figure 1.2) comes into focus to effectively operate DC systems in parallel with the AC network [2, 13, 14]. Thus, this type of network will consist of power electronics-based generations and loads.

(C) Power quality: With the advancement of technology and society, there is a need for better power quality. The integration of wind and solar power, and the use of microgrids, and growth of loads such as electric vehicles, etc., has made impacts on power quality [4]. This has led to usage of power-electronics-enabled equipment such as static var compensator (SVC), thyristor controlled reactor (TCR), thyristor control phase shifter (TCPS), thyristor switched capacitor (TSC), unified power flow controller (UPFC), unified power quality conditioner (UPQC), etc., which can be used to enhance the power quality of the power system [4, 10]. These devices are quite essential for stabilizing the power system

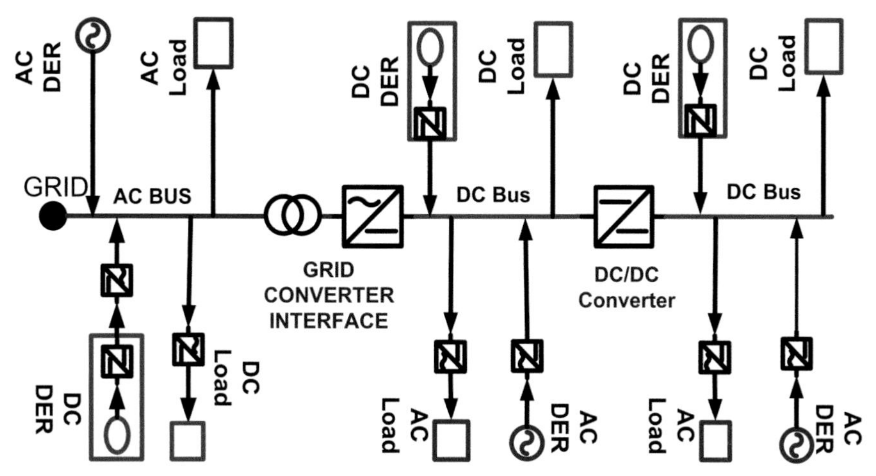

Figure 1.2 Modern power electronics enabled distribution system.

voltage, improving the power factor, and enhancing the power quality by controlling the reactive power.

(D) Smart homes, smart buildings, and smart cities [3, 15, 16]: A closely related power electronics application at the system level is the proposal for having smart buildings, smart homes, and smart cities. Smart grid is the backbone for building smart homes or cities since power will be supplied from the smart grid along with support from renewable energy sources. The scope of smart homes is wide and ever changing, but it usually embeds the following features: smart lighting, smart appliances, mobility, personalized energy, smart security, mobile communication, and remote personal care. A smart sensor is often needed for the purpose of intelligent control and monitoring. Power electronics will play an essential role and will form the backbone for modern power conversion and control intelligence for upcoming or futuristic smart cities or building [3].

(E) Storage integration [3]: The latest trend in power system integration is the development and induction of energy-storage equipment (Figure 1.3) and assets to enhance system performance, system reliability, system congestion, power availability, and hardware reliability [17]. There are many types of power storage devices being utilized viz. pumped hydro, batteries, compressed air, hydrogen, flywheels, and fuel cells. Power electronics-based converters can be best utilized for storage integrations into the system [18].

(F) Electronic loads [19, 20]: As technology evolves, the need for more efficient and accurate test instruments have increased and adapted new technologies [19]. For most electronic applications today, utilizing energy-proficient and dependable power sources is essential. Consequently, it is vital to have a test instrument that can precisely depict results that characterize power to equipment, for example, electrical vehicles, PC power supplies, 3G cells, and even batteries. Programmable DC electronic burdens are one such kind of instrument that will help with testing under different settings, setups, plans, and philosophies [21]. The purpose of this application note is to offer a general opportunity of the usage of DC loads. DC electronic loads characterize a power supply's responses to various load conditions [21]. Static switches are used by modern electronic loads. This minimizes ringing and allows management of less-than-ideal system behaviour. These days, they are recognized as general-purpose instruments that are capable of testing almost all kinds of DC power equipment including fuel cells, DC-DC converters, batteries, LED drivers, generators, and solar cells.

(G) Electrical machines, drives, and traction systems: With the advancement in cost-effective power converters, there is increased demand to use motor drives for better controllability [5]. Thus, there is an emphasis on enhancing the stability, features and performance of electric motor drive systems through proper control, employing multi-phase motors, and the pertinent modelling of electric motors over a wide loading range.

(H) Fault diagnosis, reliability, and condition monitoring [5]: To enhance overall power system longevity, fault analysis, fault-resistant control and system health management systems are of utmost importance. Power electronics engineers and researchers are working to cater for these stated requirements. Open-circuit fault diagnosis and the fault tolerance control of three-phase active rectifiers as an intrinsic stage of many power electronics applications have been addressed in [22]. The automated diagnosis system has been developed for fault detection on induction motors under transient conditions [23]. The fault-tolerant control strategy has been addressed for five-phase induction motors under four- and three-phase operation in [24]. Finally, in [25], a review report is written on a health management system for batteries (lithium-ion) particularly focused on electric vehicle applications.

Although the twenty-first century can be identified as the golden age of power electronics applications, more in-depth research and development still needs to be carried out in this area to accelerate the deployment of

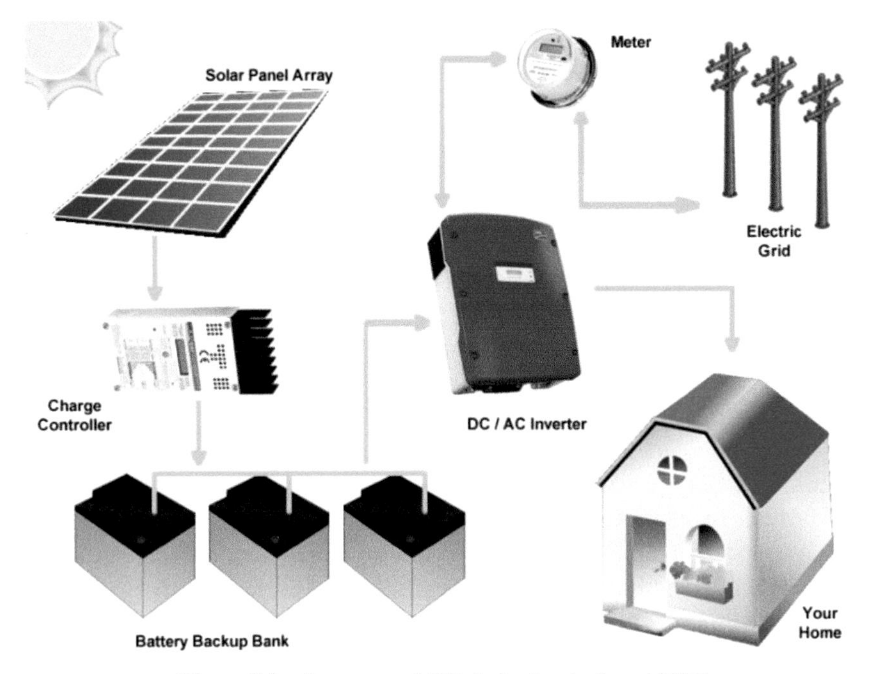

Figure 1.3 Storages and DGs induction to the grid [26].

power electronics applications. For the successful operation of such systems, optimal planning, optimal control, optimal protection, and various kinds of analyses must be carried out extensively. For the problems and their optimal solutions, the use of soft computing techniques should be a viable option. Therefore, the need of the hour is to explore the pertinent area of soft computing techniques applications to power electronics-enabled energy systems. This book focuses on technologies and solutions for the optimal planning, analysis, operation, control, and protection challenges associated with the inclusion of power electronics-enabled energy systems utilizing the concept of soft computing.

1.2 Outline of Book

Over the past 30 years, significant efforts have been made to explore the use of soft computing techniques in various engineering domains, including power systems. Motivated by the above lacuna, following chapters have been included in this book:

- Chapter 2 presents a comprehensive overview of different soft computing techniques, such as fuzzy logic, neural networks, support vector machines, and non-traditional optimization methods, as it applies to power system applications. The chapter provides detailed explanations of these techniques, including their underlying principles, to enhance the reader's comprehension and offer a holistic perspective on soft computing.
- In Chapter 3, the electrical grid is seen as undergoing a major transformation which will eventually lead to a smart grid that is based on an integrated AC-DC system. Power flow (PF) or load flow study is very essential for proper planning, analysis, and operation of a power system. Here, an AC-DC distribution load flow method is introduced. Developing a load methodology for the AC-DC distribution network is complex but relevant, especially due to the involvement of power converters. In this work, a backward-forward sweep (BFS) approach based on graph theory and matrix algebra is proposed for solving the PF for the radially configured AC-DC distribution networks with distributed energy resources (DERs). The feasibility of the proposed PF algorithm has been explored on AC-DC distribution test networks. The test results show that the approach is feasible and precise.
- Chapter 4 discusses the importance of a power flow study in understanding power system operation and its ability to supply power to loads. It covers the calculation of total system and individual line losses, determination of transformer taps settings for voltage control, and optimization of generator control settings for maximum power capacity. The study's results, including active power, reactive power, voltage magnitude, and phase angle at each bus, provide critical insights into system performance and optimization.
- The work, detailed in Chapter 5, addresses the crucial need for efficient power management in grid-connected or islanded microgrids with diverse distributed energy sources. It proposes a hybrid optimization technique, Hybrid Feedback Particle Swarm Optimization and Modified Cuckoo Search (HFPSOMCS), to enhance scheduling efficiency and cost-effectiveness. The study compares this approach with traditional Particle Swarm Optimization (PSO) and modified Cuckoo Search (MCS) algorithms, demonstrating superior performance. The developed Artificial Intelligence (AI)-based Energy Management System (EMS) effectively regulates power flow, optimizes energy production and consumption, and minimizes grid energy costs. The inclusion of machine

learning for precise solar and wind predictions and cloud computing for data analysis further enhances the system's capabilities. Overall, this work significantly advances the field of smart grid technology by improving microgrid operational efficiency and reducing maintenance costs.

- Photovoltaic (PV) systems have grown in importance over the last decade. Due to their unique designs and growing commercial interest, grid-connected single-phase transformer less (TL) solar inverters are now being considered. In Chapter 6, grid-connected inverters without transformers with power quality features are explored. This chapter describes the grid-connected solar inverters with negative, positive, and zero cycle operations. A detailed examination of all the possible topologies is carried out. The simulated results are shown to verify the effectiveness of the systems.
- With the utilization of power electronics-enabled generations and load, a lot of developments have been made in the areas of transmission and distribution systems and it has led to the development of power electronics-enabled energy systems. One of the examples, in the transmission system, is HVDC transmission. Chapter 7 provides concise information about the role and challenges of HVDC systems present in modern power systems and smart grids.
- Cyber–physical system perspective of smart grid and its security aspect is a new area of research that has attracted rapidly growing attention in the government, industry, and academia. In Chapter 8, we introduce the concept of cyber–physical system of smart grid, security concerns, the communication architecture, security requirements, security vulnerabilities, intrusion detection and cyberattack prevention, and defence approaches in the smart grid. It is also summarized that the standard and secure network protocols must be utilized to achieve efficient and secure information delivery in the smart grid.
- In Chapter 9, protection challenges during power swing and its solutions are discussed. Mathematical analysis of reduction of intensity of power swing by series compensation is provided. Various advanced techniques of fault detection such as CUSUM, TKEO-based techniques are discussed which are very effective during power swing. Fault direction estimation techniques based on superimposed negative sequence and superimposed positive sequence components are discussed and its mathematical analysis is also provided using local parameters.
- Finally, in Chapter 10, the conclusion and future scope are presented.

References

[1] Krishna Murari, N. P. Padhy and S. Kamalsadan, "Soft Computing Applications in Modern Power and Energy Systems - Select Proceedings of EPREC 2022 (ISBN: 978-981-19-8352-8)," Springer, Vol. 1, 2023.

[2] K. Murari and N. P. Padhy, "A Network-Topology-Based Approach for the Load-Flow Solution of AC-DC Distribution System with Distributed Generations," IEEE Trans. on Industrial Informatics, vol. 15, no. 3, pp. 1508-1520, March 2019.

[3] D. Tan, "Emerging System Applications and Technological Trends in Power Electronics: Power electronics is increasingly cutting across traditional boundaries," IEEE Power Electronics Magazine, vol. 2, no. 2, pp. 38-47, June 2015.

[4] Mengyao Yang, "Application of Power Electronics in Power System," International Conference on Education, Management, Computer and Medicine (EMCM 2016), China, 2016.

[5] Frede Blaabjerg, Tomislav Dragicevic, and Pooya Davari, "Applications of Power Electronics," Electronics (MDPI), Editorial, pp. 1-7, 2019.

[6] Adefarati and R. C. Bansal, "Integration of renewable distributed generators into the distribution system: A review," IET Renew. Power Gener., vol. 10, no. 7, pp. 873–884, Jul. 2016.

[7] G. F. Reed, B. M. Grainger, A. R. Sparacino, and Z. H. Mao, "Ship to grid: Medium-voltage DC concepts in theory and practice," IEEE Power Energy Mag., vol. 10, no. 6, pp. 70–79, Nov./Dec. 2012.

[8] R. A. Kaushik and N. M. Pindoriya, "A hybrid AC-DC microgrid: Opportunities & key issues in implementation," 2014 International Conference on Green Computing Communication and Electrical Engineering (ICGCCEE), Coimbatore, India, 2014, pp. 1-6.

[9] N. Flourentzou, V. G. Agelidis and G. D. Demetriades, "VSC-Based HVDC Power Transmission Systems: An Overview," IEEE Trans. on Power Electronics, vol. 24, no. 3, pp. 592-602, March 2009.

[10] V. K. Sood, HVDC and FACTS Controllers: Applications of Static Converters in Power Systems. Boston, MA: Kluwer, 2004.

[11] M. P. Bahrman, "HVDC transmission overview," 2008 IEEE/PES Transmission and Distribution Conference and Exposition, Chicago, IL, USA, 2008, pp. 1-7.

[12] D. J. Hammerstrom, "AC Versus DC Distribution Systems Did We Get It Right?" 2007 IEEE Power Engineering Society General Meeting, Tampa, FL, USA, 2007, pp. 1-5.

[13] Krishna Murari, Narayana Prasad Padhy and Sukumar Kamalasadan, "Backward-Forward Sweep Based Power Flow Algorithm for Radial and Meshed AC-DC Distribution System," 2021 IEEE Industry Applications Society Annual Meeting, Vancouver, Canada, 2021.
[14] Krishna Murari and Narayana Prasad Padhy, "An Efficient Load Flow Algorithm for AC-DC Distribution Systems," Electric Power Component and Systems (Taylor & Francis), Vol.46, No.8, pp. 919-937, 2018.
[15] A. Chatterjee, S. Paul, and B. Ganguly, "Real Time Multi-Objective Energy Management of a Smart Home," 2020 IEEE International Conference on Power Electronics, Drives and Energy Systems (PEDES), Jaipur, India, 2020, pp. 1-6.
[16] M. Manic, D. Wijayasekara, K. Amarasinghe and J. J. Rodriguez-Andina, "Building energy management systems: the age of intelligent and adaptive buildings," IEEE Ind. Electron. Mag., vol. 10, no. 1, pp. 25-39, 2016.
[17] H. Johal, D. Manz, K. O'Brien and J. Kern, "Grid integration of energy storage," 2011 IEEE Power and Energy Society General Meeting, Detroit, MI, USA, 2011, pp. 1-2.
[18] M. Kumar, K. Shanmugam, K. Pradeep and M. Filippone, "Grid integration and application of Battery Energy Storage Systems," 2022 IEEE International Conference on Electronics, Computing and Communication Technologies (CONECCT), Bangalore, India, 2022, pp. 1-6.
[19] https://www.keysight.com/us/en/assets/7018-06481/white-papers/5992-3625.pdf
[20] J. Peng, Y. Chen, Y. Fang, and S. Jia, "Design of Programmable DC Electronic Load," 2016 International Conference on Industrial Informatics - Computing Technology, Intelligent Technology, Industrial Information Integration (ICIICII), Wuhan, China, 2016, pp. 351-355.
[21] L. F. Serna-Montoya, J. B. Cano-Quintero, N. Muñoz-Galeano and J. M. López-Lezama, "Programmable Electronic AC Loads: A Review on Hardware Topologies," 2019 IEEE Workshop on Power Electronics and Power Quality Applications (PEPQA), Manizales, Colombia, 2019.
[22] H. Cheng, W. Chen, C. Wang, and J. Deng, "Open Circuit Fault Diagnosis and Fault Tolerance of Three-Phase Bridgeless Rectifier," Electronics, vol. 7, no. 11, p. 291, Nov. 2018.

[23] J. Burriel-Valencia et al., "Automatic Fault Diagnostic System for Induction Motors under Transient Regime Optimized with Expert Systems," Electronics, vol. 8, no. 1, p. 6, Dec. 2018.
[24] S. C. Rangari, H. M. Suryawanshi, and M. Renge, "New Fault-Tolerant Control Strategy of Five-Phase Induction Motor with Four-Phase and Three-Phase Modes of Operation," Electronics, vol. 7, no. 9, p. 159, Aug. 2018.
[25] Z. Omariba, L. Zhang, and D. Sun, "Review on Health Management System for Lithium-Ion Batteries of Electric Vehicles," Electronics, vol. 7, no. 5, p. 72, May 2018.
[26] https://www.tuppersteam.com/blog/First-Came-Solar-SystemsBattery-Storage-Is-on-the-Horizon

2

Soft Computing Techniques and Their Application in Power Systems

Anupam Kumar

National Institute of Technology, Patna, Ashok Rajpath, Bihar, PIN - 800005
Email: anuanu1616@gmail.com

Abstract

In the past three dccades, extensive research and development works have been reported on soft computing techniques for many engineering fields, including power system applications. This chapter presents different soft computing techniques such as fuzzy logic, neural networks, support vector machine, and non-traditional optimization techniques. The chapter provides details of these techniques, including their working principles.

Keywords: Soft computing, neural network, fuzzy logic, state vector machine, genetic algorithm, particle swarm optimization, ant colony optimization

2.1 Introduction

The applications of soft computing techniques have gained popularity over past years and are influencing all pertaining fields of science and engineering. One of the relevant fields is power systems, in which solutions for optimal planning, analysis, operation, and control have been greatly influenced using various soft computing approaches. These techniques offer an effective solution for studying and modelling stochastic behaviour, imprecision, uncertainty, decision-making, partial truth, and approximation problems, making them more valuable than traditional techniques. The soft computing term was introduced by Professor Lofti A Zadeh [1] in 1965. After many years, around 1995, he also introduced the term hard computing [2], which means providing precise/unambiguous results and is ideal for solving problems that have well-defined mathematical solutions. It is based on workings of the human mind, biological behaviour etc., and is broadly known as artificial intelligence.

Some important branches of soft computing techniques are neural networks, fuzzy logic, evolutionary computation, probabilistic reasoning, etc. A neural network can learn and adapt as per requirements. Fuzzy logic is based on an expert's knowledge and evolutionary computation is established on Darwin's principle of "survival of the fittest strategy" and genetic evolution. The soft computing techniques are highly robust, adaptable, and low cost. Furthermore, it is observed that soft computing techniques are also applicable in different fields such as robotics [3, 4], aerospace [5], pattern classification and recognition [6], medical applications [7, 8], market prediction, power system applications [9–11], etc.

This chapter starts with an introduction to soft computing. A brief discussion on different soft computing techniques is provided in Section 2.2. This is extended in sections 2.3, 2.4, 2.5, and 2.6 with discussion on different soft computing approaches such as fuzzy logic systems, artificial neural networks, various non-traditional optimization techniques, and state vector machines, respectively.

2.2 Soft Computing Techniques

In power system applications, soft computing techniques play a crucial role for controlling, identification, predicting, etc. The role model of soft computing techniques is the human mind that can argue, learn from experience, and take human-like decisions. As shown in Figure 2.1, the principal member of soft computing technique comprises fuzzy logic systems (FLS), artificial neural networks (ANNs), nontraditional optimization techniques, machine learning, etc. The terminology of soft computing was first introduced in [1] with the idea of fuzzy logic [6]. Presently, soft computing has enlarged its coverage to include swarm intelligence, artificial life, ant behaviours, bio-inspired, chaos theory, learning theory, etc. Soft computing techniques have recently become a well-known tool for solving many engineering problems; for example, they have been used in many areas, namely electrical, electronics, computer engineering, and medical fields. A brief discussion on different soft computing approaches is discussed in the next section.

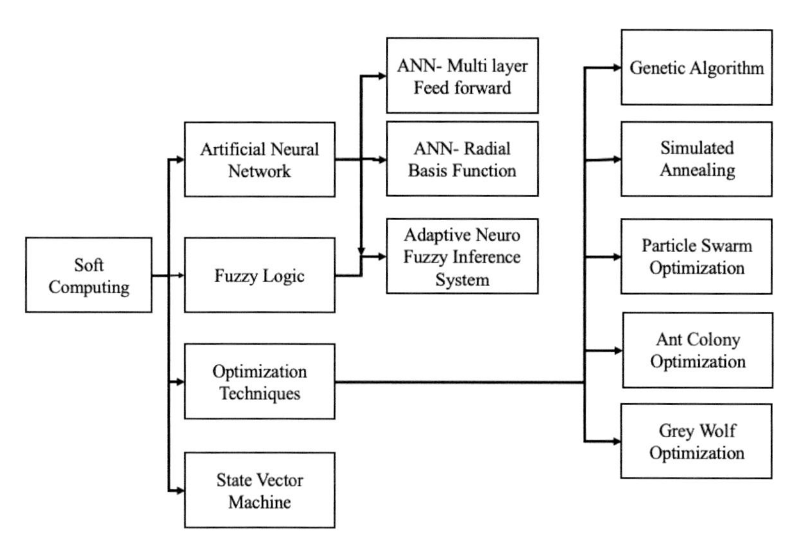

Figure 2.1 Principal members of soft computing techniques.

2.3 Fuzzy Logic System

Fuzzy logic system (FLS) is based on fuzzy sets theory which is an extension of multi-valued conventional logic. Fuzzy logic deals with approximate rather than precise modes of reasoning. It is sort of computational model which provides a mathematical tool for handling various type of uncertainties. It is a decision-making process to predict and diagnose in a fuzzy logic framework that is structured as a set of if $<$ situation $>$ then $<$ action $>$ rules where both situation and action have suitable fuzzy representation. The fuzzy logic framework is used in different kinds of power system applications [7] and an expert's knowledge is represented on behalf of working for control action. In general, FLSs are mostly recommended for systems which are complex and mathematically ill-defined. Thus, the performance of an FLS depends on human expertise about the system and the knowledge acquisition techniques to convert human expertise to appropriate fuzzy if-then rules as well as a proper fuzzy membership function for each fuzzy variable.

This subsection provides a brief description and basic terminology of FLSs. The FLSs comprise of four core essential blocks: fuzzification, fuzzy inference, fuzzy rule base, and defuzzification, as depicted in the dotted box in Figure 2.2. Apart from these blocks, two more blocks are also used for converting them in actual range, namely pre-processing and post-processing blocks. A brief description of the various FLS components is given below:

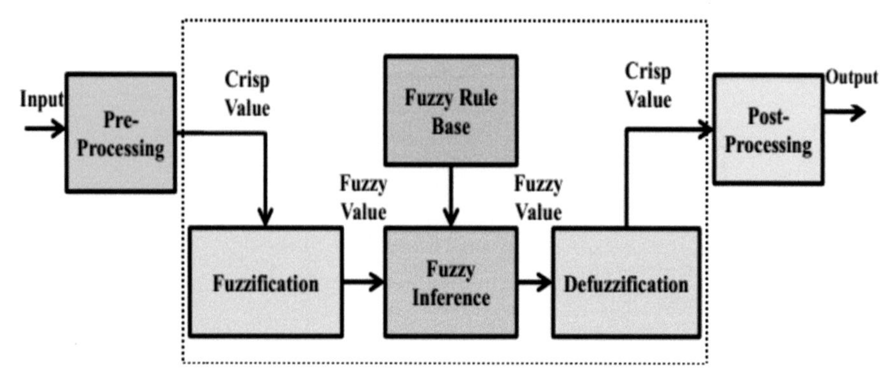

Figure 2.2 Basic blocks of FLSs.

2.3.1 Fuzzification

Fuzzification blocks have an essential role to convert the crisp input into fuzzy values or linguistic values. The input and output values of the FLSs can be distinguished by various membership functions (MFs). Usually, the selection of MFs is based on symptoms of the input data.

2.3.2 Rule base and inference mechanism

Rule base is an essential component of FLS design that depends on expert knowledge of plants and type of plant. It can be easily generated by human experience for given diseases symptoms. The fuzzy inference engine is employed to obtain a suitable fuzzy output action based on the allocation of each rule. Typically, Mamdani and Takagi-Sugeno techniques are used as fuzzy inference in the literature.

2.3.3 Defuzzification

Defuzzification task is the procedure of producing a compatible result in FLS. It converts fuzzy sets into crisp sets that are feed to a controlling plant. The center of gravity procedure is applied in this work to defuzzify the inference output.

2.4 Artificial Neural Networks (ANNs)

In simple terms, artificial neural networks (ANNs) are a network of neurons. The foundation of ANNs is inspired from the working of a neuron in the human brain. The first ANN was designed by psychologist Frank Rosenblatt [12] in 1958 and is known as single-layer perceptron. Perceptron models were developed to learn how the human brain visualizes data and learn to recognize objects. In addition, a human brain has a highly complex nonlinear and parallel computer, which can organize its constituent structural elements. Its structural constituents are known as neurons. The neurons are interconnected not in a straightforward way but in a rather complex way, so complex that many things are still unknown. There is a connection between neurons and neurons, which is how a network is realized. And this network is highly complex, as well as nonlinear and massively parallel. Similar to how the human brain uses neurons to transmit the signals from different parts of the body to the brain, ANN also uses neurons to process the information and send it between its multiple layers. ANN is a powerful tool to not only process, know patterns but also detect the unknown parts of information and hence make sense of them. The detection and processing of the unknown information in an ANN is called network learning. This is done by mathematically adjusting the values, or weights, of the neurons in the network until the most optimized and efficient results are obtained. Neural networks can be trained when the desired output is not known. The neural networks are provided with some patterns or group of facts and then are left to learn independently and finally reach a stable state after many iterations. This kind of neural network training is called unsupervised learning [13]. Furthermore, the other type of learning known as supervised learning or multilayer feed-forward neural networks is based on input/output relations. In this type, the neural network is provided with the desired output and the input patterns, and it is required to adjust the weights of neurons such that the calculated output comes out to be closest to the desired output. The multilayer feed-forward neural networks, abbreviated as MLF neural networks, are the most popularly used neural networks as shown in Figure 2.3 [14].

In addition, the learning of neural networks rapidly increased after successfully implementing the backpropagation (BP) algorithm on a computer network [15]. Seppo Linnainmaa in 1970 [16] implemented the first BP algorithm. Paul Werbos in 1974 [17] analyzed that BP algorithms can be used in neural networks to improve their performance. In their research, they discussed the theory of BP with all required mathematical functions and assumptions. BP network uses different learning algorithms to generalize BP algorithm for a gas market and show it can be generalized to any type of neural network [18]. A common learning algorithm is gradient descent (GD) which is based on BP and differential calculus, and it is used to find local minimum of a function. Partial derivative of error, with respect to each variable, is calculated using the chain rule (Leibnitz, L'hopital in 1667) then

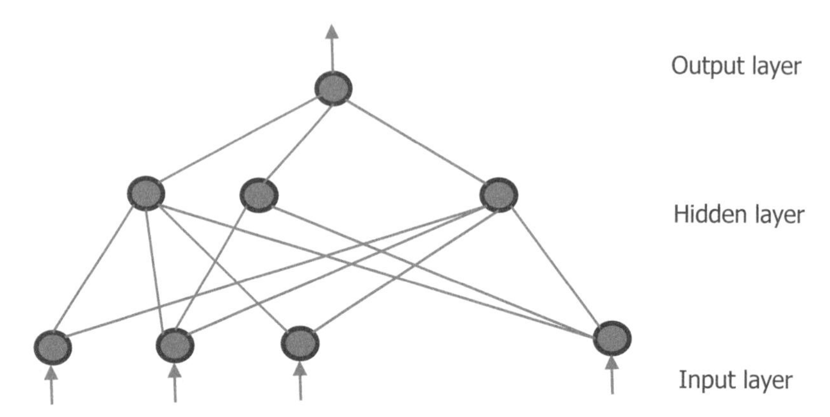

Figure 2.3 A basic feed-forward ANN having three layers.

it is minimized. GD algorithms have their advantages and disadvantages. The authors in [19] discuss the limitations of GD such as overshooting, vanishing gradient. GD never ensures the most optimal solution; it only provides the local optimal solution. Nowadays, ANNs are found in many applications such as finance, surgery, computer vision problem, power system applications, etc.

2.4.1 MLF neural networks

Neural networks are networks of neurons which are hugely interconnected with each other. NNs consist of a neuron as its fundamental unit of the network. A neuron receives input from the previous layer, applies some activation function on it, transmits it as an input to the next layer or generates an output. Figure 2.4 depicts a typical 2-input, 1-output, 3-layer ANN which generally has three types of layers: input, output, and hidden layer.

A brief description of each of the layers along with activation function, bias, and other techniques to minimize the error is provided below:

Input layer: The inputs are supplied at the input layer and these inputs are then multiplied with corresponding weights and added together to feed the neuron as input to the next layer. Here x_1, x_2 are inputs for the ANN.

Weights: It is the parameter within a neural network that weights connecting the input data to the hidden layers of the network. These weights are adjusted while applying learning technique such as backpropagation to minimize the error. As shown in the Figure 2.4, w_1, w_2, w_3, w_4, w_5, w_6 are weights of the neurons.

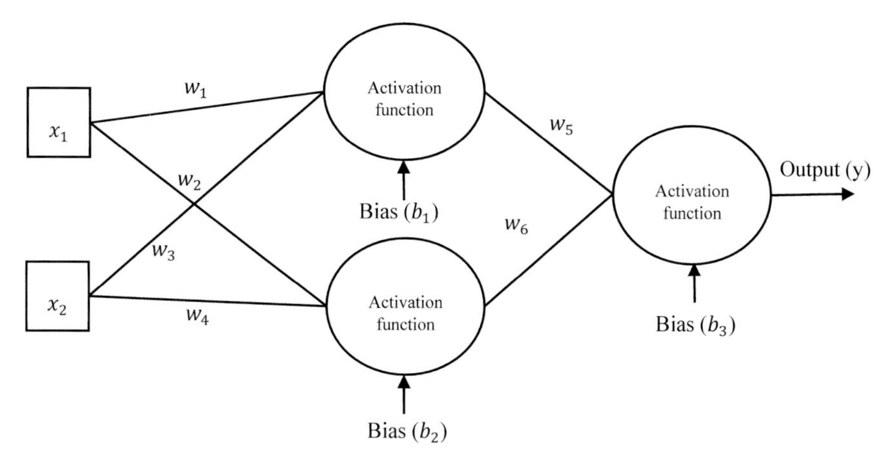

Figure 2.4 A neural network with an input layer having 2 inputs, output layer with 1 output and 1 hidden layer.

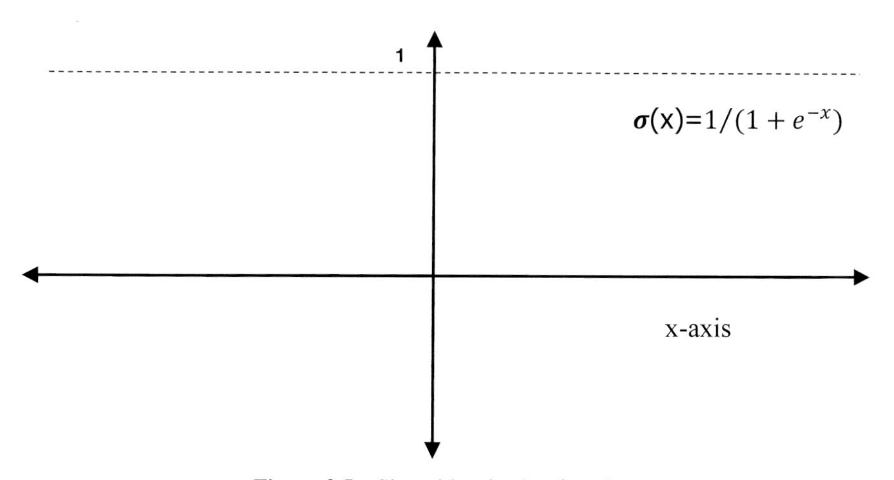

Figure 2.5 Sigmoid activation function.

Activation function: It is the function that applies a nonlinear transformation on a given input before sending it to the next neuron or sending it as an output. Most widely used activation functions are sigmoid (as shown in Figure 2.5), rectified linear activation (RELU), etc.

$$\text{Sigmoid function, } \sigma(x) = \frac{1}{1 + e^{-x}} \tag{2.1}$$

Bias: This factor lets us relocate the activation function by a constant value. Biases (i.e., b_1, b_2, b_3) are needed to avoid overfitting of the networks, as shown in Figure 2.4.

Output layer: It is the last layer of the neural network where the output of the system is generated.

Hidden layers: All layers between input and output layers are known as hidden layers. Increasing the number of hidden layers increases accuracy and learning of the algorithms, but also increases the complexity which may lead to extra cost or extra computational time.

Assume *y* is the output of the ANN and a sigmoidal function $\sigma(x)$ is used on the neurons, the ANN can be mathematically modelled as [16]:

$$y = \sigma\left\{w_5\left(\sigma(x_1 w_1 + x_2 w_3) + b_1\right) + w_6(\sigma\left(x_1 w_2 + x_2 w_4\right) + b_2)\right\} + b_3 \tag{2.2}$$

Backpropagation: This is one of the common learning techniques for ANN and its required error function. Some popular error functions are root mean square error (RMSE), log loss, etc. The RMSE is the average of square root of errors and given as:

$$\text{Error} = |\text{Actual} - \text{Predicted}|. \tag{2.3}$$

Moreover, once the error is calculated, it is then differentiated with respect to each weight, bias with help of the chain rule. These weights are then updated accordingly. This phenomenon is known as backpropagation [6]. In this process, the rate of change of error is calculated with respect to a particular weight and this $weight$ is updated as in [6]

$$\text{weight} - \text{learning rate} \cdot \text{rate of change} \tag{2.4}$$

2.4.2 Radial basis function

Radial basis function (RBF) model was introduced by Broomhead and Lowe in 1988 [21]. The RBF neural network is a type of ANN that is generally used for function approximation problems; it uses radial basis functions as activation functions. It is composed of three layers: the input layer, the hidden layer, and the output layer. Here, the activation functions are typically implemented with Gaussian functions [22]. When compared to the MLP, it has a much faster training process; therefore, it can be used in place of an

MLP network. RBF neural network has caught the interest of scientists and engineers as its network structure is simple, has faster learning speed, and superior approximation capabilities [20]. The general architecture of RBF neural network is depicted in Figure 2.6 [20]. Furthermore, the output of the RBFNN is given by eqn (2.5) [20]:

$$\mathrm{y}\,(x) = \sum_{j=1}^{k} w_{ij}\varphi\left(||x - c_j||\right) \tag{2.5}$$

where

x – Input parameter,

y – Output of the network,

k – Total number of neurons in the hidden layer,

c_j – Centre of the *j*th hidden node,

w_{ij} – Denotes weight of the connection from jth neuron in the hidden layer to the ith neuron in the output layer,

$|| \; ||$ – Denotes Euclidian distance, and

φ – Denotes radial basis function.

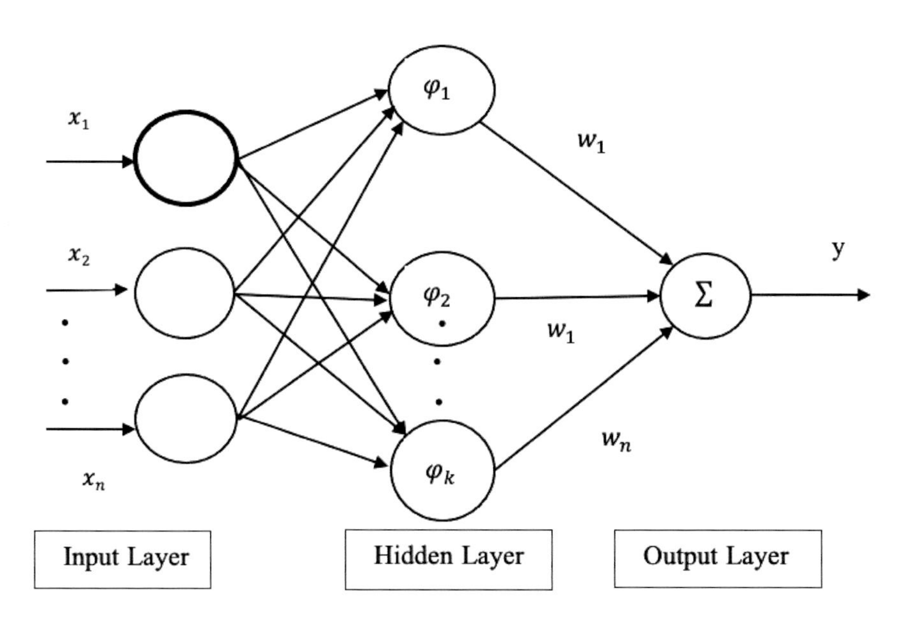

Figure 2.6 Architecture of radial basis function neural network [20].

There are so many kinds of radial basis functions; the most widely used RBF is the Gaussian function which is given in eqn (2.6) [20]

$$\varphi\left(||x - c_j||\right) = e^{\left(-\frac{||x-c_j||^2}{2\sigma_j^2}\right)}. \tag{2.6}$$

In eqn (2.6), σ_j denotes the width of the jth hidden neuron.

The main differences between RBF and MLP neural networks [23] are as follows:

- RBF neural network contains only one hidden layer, whereas several hidden layers are present in MLP network.
- The hidden layer in an RBF network is nonlinear, while the output layer is linear. On the other hand, the multilayer network has nonlinear layers throughout.
- A RBF's activation function is determined by calculating the distance between the pattern and the sample. A multilayer network's activation function is based on the calculation of an inner product.
- In MLP network, all nodes have the same activation function but in case of RBF networks, it is not necessary to have same activation function in all the nodes.

2.4.3 Adaptive neuro-fuzzy inference system (ANFIS)

The soft computing domain included a wide range of models based on learning capacity, flexibility, complexity, and scalability. Therefore, the advantages of FLS and learning capability of ANN are combined to develop ANFIS for addressing their limitations while also providing robustness and powerful prediction capabilities. The ANFIS is a network which is implemented using the integration of both fuzzy systems and neural networks. It captures the merits of both in a unique way by providing great estimation accuracy such as mean magnitude of relative error and high prediction. Being a universal estimator, ANFIS employs if-then rules and has the ability to mimic exceedingly complicated, nonlinear systems through learning. The ANFIS design is based on the widely used gradient descent technique, which employs the backpropagation algorithm to calculate gradients and least square estimation to locate coefficients of errors for optimizing ANFIS parameters. An ANFIS was developed in the early 1990s [24, 25]. The structure of the ANFIS, (Figure 2.7), consists of five layers where input and output nodes represent the input states and output response, respectively. In addition, the middle

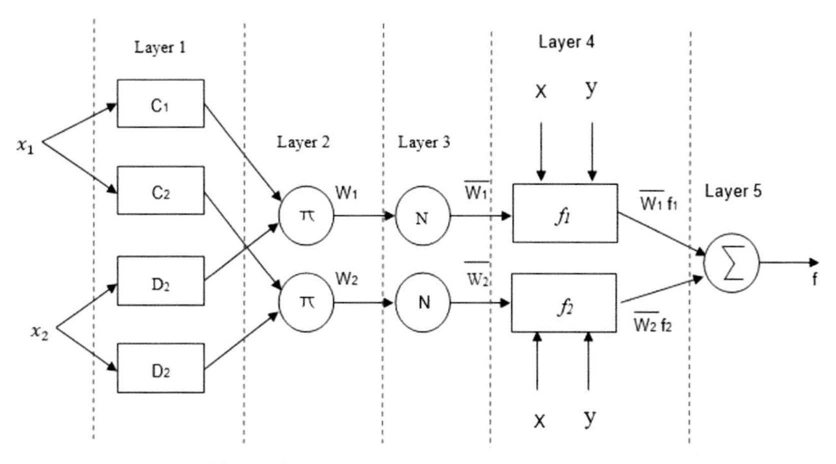

Figure 2.7 Structure of the ANFIS model.

layer nodes contain membership functions and rules. Here, the two types of nodes in the ANFIS 5-layers architecture are present i.e., fixed and adaptable [26, 27]. Now, the basic mathematics and working are discussed as below:

As shown in Figure 2.7, Layer 1 gives the output (O_i^1) as:

$$O_i^1 = \mu_{\mathrm{C}i}(\mathrm{x}), i = 1, 2, 3, \ldots O_i^1 = \mu_{\mathrm{D}i}(\mathrm{y}), i = 1, 2, 3, \ldots \tag{2.7}$$

where μ_{i1} (x) and μ_{i2} (y) are the membership functions of an adaptive node. O_i^1 represents the output of the first layer and "*i*" represents the number of the membership function.

Layer 2 quantifies the firing strength with the help of product operations on a given set of input values:

$$O_i^2 = w_i = \mu_{\mathrm{C}i}(x) \times \mu_{\mathrm{D}i}(\mathrm{y}), i = 1, 2, 3, \ldots. \tag{2.8}$$

Layer 3 plays the role of normalizing the strength of rule from previous layer output using,

$$O_i^3 = \underline{w}_i = w_i / \Sigma w_i, i = 1, 2, 3, \ldots \tag{2.9}$$

Layer 4 nodes are adaptive and handle fuzzy rule operation with linear coefficients of consequent nodes.

$$O_i^4 = \underline{w}_i \cdot f_{\mathrm{i}} = \underline{w}_i \cdot ((m_a)\, x + (n_a)\, y + s_a)\,, i = 1, 2, 3, \ldots \tag{2.10}$$

where m_a, n_a, and s_a are linear parameters and $\underline{w}_i$ is the output of layer 3.

Layer 5 performs the last part of the process i.e., defuzzification of the consequent set providing results of rules.

$$O_i^5 = \sum_{i=1}^{n} \underline{wi} \cdot f_{\mathrm{i}}. \tag{2.11}$$

The ANFIS is being applied to different areas of interests such as power system [28], forecasting, economy, environment, because of its ability to capture nonlinear part of process. Moreover, it intelligently handles all the drawbacks of ANNs by combining both neural networks and fuzzy logic systems in an efficient manner.

2.5 Support Vector Machine (SVM)

Support vector machine (SVM) was introduced by Cortes and Vapnik to train line equipment effectively [29]. It is a (supervised) machine learning (ML) algorithm and is utilized for editing and regression challenges. It is widely utilized in classification problems. One of the simplest functions of such models is the linear separation of different vectors representing two classes. It finds a hyperplane with a single size less than the actual size of the vectors dividing into two phases. The purpose of the separation is to increase the space of class elements from the hyperplane to its various locations. The closest class elements from these two classes are termed support vectors. SVMs are developed and expanded to handle more complex tasks, e.g., the division of multiple classes even when classes are not linearly divided. So, the goal of the SVM is to minimize the number of incorrectly classified elements and to increase the distance between the dividing hyperplane and the supporting vectors. Support vectors thus provide addresses to each monitoring [27]. Furthermore, there are four main ideas that cover the basics of the algorithm:

Max-margin classifier:

In ML, it is a separator category that can provide a related distance from the decision boundary for each model. A consideration section that best shows how SVM works. It validates the view of the best line separator.

Lagrangian multipliers:

It is a strategy to find the size of the area and the minimum function of the work under equity limits such that under the condition that one or more numbers must be directly satisfied with the selected variance values. So, it is a way to make problems simpler to solve.

The kernel function:

It is used to generate data in a high-resolution area for distribution that is linear. It is one of the mathematical functions used by SVM. Types of kernels are linear, nonlinear, polynomial, radial basis function, and sigmoid.

Complexity:

It depends on the training set count. The SVM complexity is O(n3), and the time complex is shown to be faster for C++ than Java, both in training and testing.

In literature, various types of SVM algorithms are presented [30]; a few of them are discussed next:

1. **C-SVM:**

Parameter C is an error name that is a penalty parameter. It can be considered as the level of separation that the algorithm must meet or the level of configuration that the SVM must meet. With large C values, there is no way the SVM processor can distinguish any single point.

2. **nu-SVM:**

The nu-SVM uses the nu parameter instead of C (used as a hyperparameter in the SVM line) as a hyperparameter to punish the wrong categories. Scope indicates the upper and lower limits where our hyperparameter can take its value.

3. **Epsilon-SVM regression:**

The epsilon (ϵ) tells the tolerance limit when no penalty is given for errors. The larger the epsilon, the more errors occur in the solution.

4. **nu-SVM regression:**

In nu SVM regression, the nu is used to determine a fraction of the support value vectors wishing to keep to the solution with respect to the total number of samples in the database.

2.6 Nontraditional Optimization Techniques

Nowadays, optimization techniques are necessary for every area where the global optimum solution is required. The optimization techniques, based on development in nature, social behaviour, physical science-based etc, have become very popular. In the literature, the various optimization techniques are reported and used by researchers, scientists, and engineers to optimize their problems efficiently for many applications including power systems. Therefore, the few well-known nontraditional optimization techniques are discussed in brief in this subsection.

2.6.1 Genetic algorithm

A genetic algorithm proposed by Holland and his collaborators in the 1960s and 1970s [31] is an intuitive search inspired by Darwin's evolution theory. Later, it was broadly explored by Goldberg [32] in his research. It mimics natural selection wherein the individuals of an arbitrary generation are selected for reproduction with the idea to obtain superior/desirable offspring for the subsequent generations. In a more general sense, a genetic algorithm is a powerful tool that can be employed to crack constrained and unconstrained optimization problems. It frequently changes a population to pick individuals from the parents and use them to produce the next offspring. Over iterations, the population "grows/evolves" towards the optimal solution. Moreover, a genetic algorithm can be utilized to solve optimization problems not well suited for standard optimization algorithms like gradient descent that struggle to optimize nondifferentiable objective functions. A genetic algorithm is generally used to optimize discontinuous, nondifferentiable, stochastic, or highly nonlinear objective functions. Mathematical optimization methods are remarkable in their own implementation, but they are restricted when the number of governing parameters increases. That's where genetic algorithms shine. In cases where data is abundant but mapping an equation to explain the data is difficult to find, genetic algorithms can be utilized to start with a random expression and evolve it towards the optimal solutions. The standard basic operators of the genetic algorithm are discussed as follows:

2.6.1.1 Initial population

A genetic algorithm begins by acquiring a group of candidate individuals to create a population of compatible solutions. Here, this population is generated randomly. The number of unique chromosomes/solutions depends on the problem, thus there is no fixed population size for every optimization problem. The smart thing that can be done with the starting population is generating the random generator in such a way that the population parameters should lie around the probable solution. Every chromosome (individual solutions) is represented by a set of parameters (variables) known as genes. Genes can take the form of bits; thus, the chromosome becomes a string of bits.

2.6.1.2 Fitness function

Fitness function employed to calculate fitness of the individual solutions that is used to segregate the population. It is the rule that determines if an individual chromosome is worthy enough to pass its traits to the following generations. The fitness function is crucial as it's the driving force that pushes the population to converge towards the solution.

2.6.1.3 Selection

In the selection process, the individual chromosomes that pass the fitness function criteria are collected in the mating pool. Two of best solutions are picked and allowed to pass their genes to make up the chromosome of the next individual solutions. The picked chromosomes are participated in the crossover (see section 5.1.4) for generating offspring. The selection process can become computationally expensive since calculating the fitness of each solution is too time-consuming. Hence, selection methods are heavily researched. Some of the well-known selection schemas are roulette wheel and tournament selection methods.

2.6.1.4 Crossover

Crossover is the most important step in a genetic algorithm. For each pair of parents to be coupled, an arbitrary crossover point is selected from within the genes. Offspring are created by swapping the genes of parents among themselves until the crossover point is reached. The latest offspring are added to the population. The offspring/child solution that is obtained shares parameters with its parents.

2.6.1.5 Mutation

In a few offspring, some of the genes can be subjected to mutation with a random likelihood. This means that some of the bits in the binary number/alphabet can be switched. A mutation is desirable to sustain diversity in the population and discourage early convergence.

Furthermore, the working principle involves certain standard steps, as shown in Figure 2.8. It is observed that the starting point is a set of individuals that can all be candidate solutions for the given problem. This set of individuals is a population and is generated randomly. Each individual can commonly be represented as a string of bits that represents a possible solution. Each bit is analogous to the genes in living things and the set of bits, or the string corresponds to a chromosome. This chromosome represents the solution, as mentioned earlier. The values of genes/bits are altered through

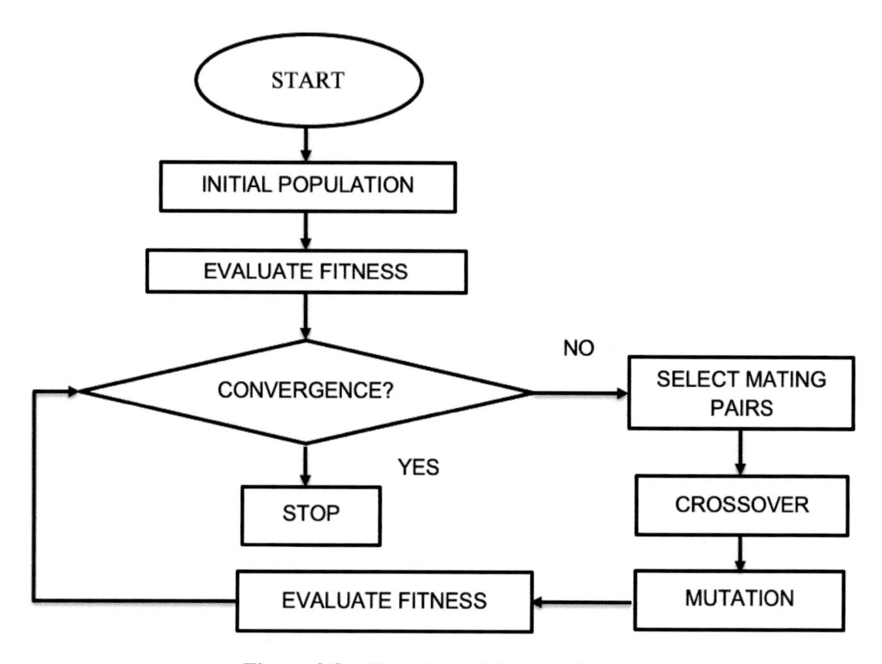

Figure 2.8 Flowchart of GA algorithm.

the generations, although the size of the population pool is kept the same. At each generation/iteration, the viability of an individual is determined by a fitness function. The fitness value is important because it determines which individuals get to forward their genes to the next generation. This is what makes genetic algorithms like Darwin's theory of evolution – mimicking the axiom of survival of the fittest.

The chromosomes of the individuals selected are transferred to the next generation in a probabilistic manner. The individuals that are deemed fit randomly mate and their genes crossover to create chromosomes of individuals in the subsequent generation. When this process happens, the mutation phenomena can happen randomly. This can increase diversity and prevent premature convergence causing inconsistent solutions. As the fittest chromosomes are selected, the average fitness value of the individuals in the following generation is always higher in comparison to the predecessors. This is how the genetic algorithm advances/evolves towards the ideal solution. These steps are repeated until the process reaches close enough to the solution. Commonly, the algorithm terminates when a maximum number of generations has been reached, assigned time has been over or a satisfactory fitness level has been reached. This condition can be considered for convergence of the algorithms.

2.6.2 Particle swarm optimization

As we know, evolutionary computation techniques are based on mimicking the natural phenomena or the process, on the similar way, particle swarm optimization (PSO) is motivated from foraging and social behaviour of a swarm. The PSO algorithm was proposed by Eberhart and Kennedy in 1995 [33] based on the natural phenomena of a school of fish, flock of birds, etc. It is also known as swarm intelligence. Authors in [33] demonstrated PSO algorithm as an optimizer for a continuous nonlinear function; moreover, this PSO technique is a direct search method which does not need any gradient information. A PSO is found to be computationally inexpensive both in terms of memory and speed and can be easily implemented. These are the main advantages of PSO since the operations are primitive concepts. Since the PSO does not involve probability distribution it makes the calculation simpler. The general flow diagram of the PSO is illustrated in Figure 2.9.

PSO is initialized with randomly generated population of particles (initial swarm). A random velocity is assigned to each particle that propagates the particle in search space towards optima over several iterations.

The PSO is an iterative process while improving a candidate solution (known as particle) with given fitness quality. These particles use a simple mathematical formula to exploit their positions and velocity to achieve an optimal result. Here, the best position of the particle in the search space is updated as per neighbourhood's best positions. The mathematical formulations for positions and velocity are given as follow:

Here, the individual position of particle (i) is adjusted as:

$$x_i^{(t+1)} = x_i^{(t)} + v_i^{(t+1)} \tag{2.12}$$

where velocity of particle is updated as follows:

$$v_i^{(t+1)} = wv_i^{(t)} + c_1 r_1 \left(p_{(i,lb)}^{(t)} - x_i^{(t)}\right) + c_2 r_2 \left(p_{(gb)}^{(t)} - x_i^{(t)}\right). \tag{2.13}$$

Here, i is the ith particle, t is the generation counter, w adds to the inertia of the particle, c_1 and c_2 are the acceleration coefficients, r_1 and r_2 are random numbers and it belongs to the range of 0 to 1, $p_{(i,\, lb)}^{(t)}$ is the local best of i-th particle, $p_{(gb)}^{(t)}$ is global best.

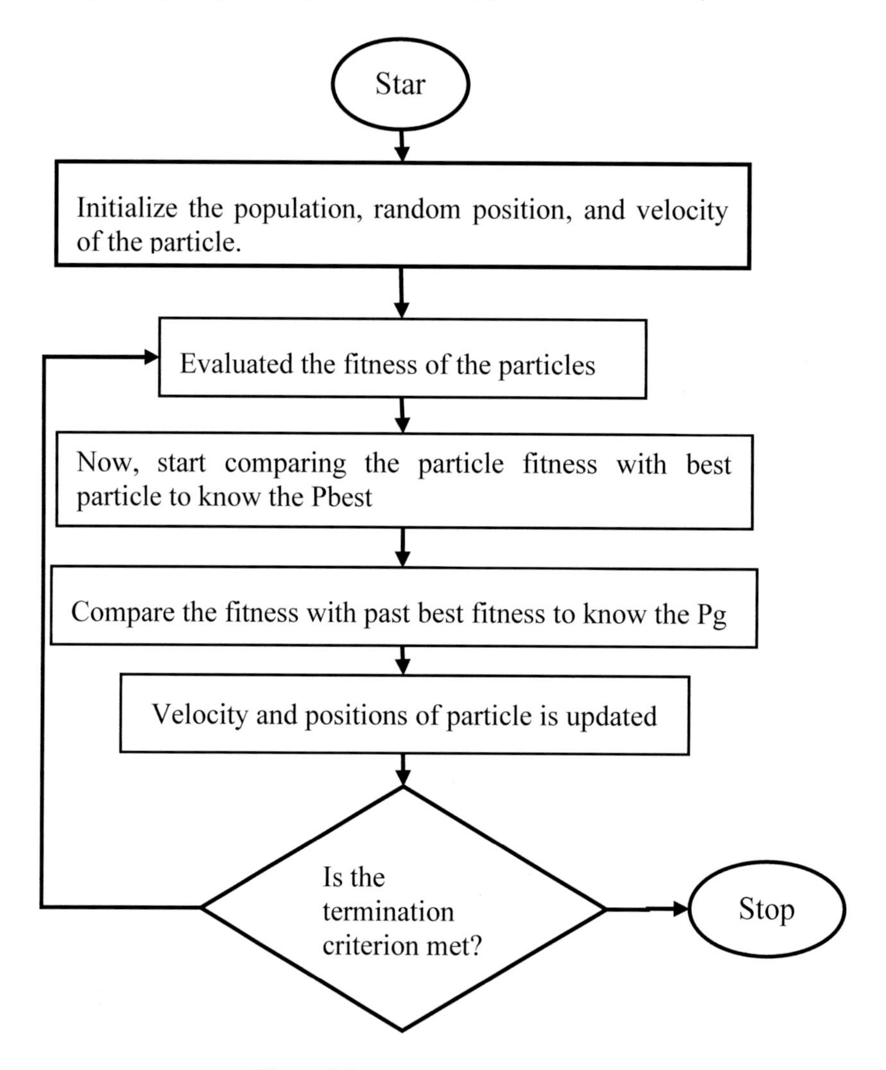

Figure 2.9 Flow diagram of the PSO.

The local and global best positions of the particles are given as:

$$p_{(i,lb)}^{(t+1)} = \begin{cases} p_{(i,lb)}^{(t+1)} = x_i^{(t+1)} & \text{if f}\left(x_i^{(t+1)}\right) < \text{f}\left(p_{(i,lb)}^{(t)}\right) \\ p_{(i,lb)}^{(t)} & \text{otherwise} \end{cases} \tag{2.14}$$

$p_{(i,\, lb)}^{(t)}$ is the personal best position of ith particle in t generation, assume minimization problem.

$p_{(gb)}^{(t)}$ is the global best position in t generation which is calculated as:

$$\text{Min}\left\{\text{f}\left(p_{(1,lb)}^{(t)}\right),\ldots\ldots\ldots,\text{f}\left(p_{(N,\text{ best })}^{(t)}\right)\right\} \tag{2.15}$$

Here, N is the number of particles in the swarm.

2.6.3 Ant colony optimization

Ant colony optimization (ACO), encouraged by nature, is a kind of meta-heuristic algorithm. It is developed to solve some of the hard combinatorial optimization problems. The ACO is based on natural ant behaviour, team planning and organization, integration to find the best solution, and storage of data for each ant [35–37]. This algorithm was proposed by Marco Dorigo in 1992 and it is inspired by a real ant's ability to find the shortest path between food and nest [34]. The ants hardly use any vision to navigate their routes and rather use the pheromone trails that are left on the ground. The pheromone trail is a chemical substance which is used as a signal for other ants. This process is called stigmergy.

The ants follow the pheromones trail left behind by other ants that have already travelled to find the food, as shown in Figure 2.10. If there are multiple paths to follow, then the ant might choose any one of the paths randomly. While travelling on any of the chosen paths, the ants leave behind a pheromone trail which strengthens the trail. As the higher number of ants follow any single trail, the stronger the trail becomes which increases its probability of being chosen by other ants. It is also important to understand that the pheromone concentration decays with time. As a result, the pheromone build up is more on shorter paths since it has less time to decay. This leads to further strengthening of the shorter paths with the shortest path being the most highlighted. This whole process is shown in Figure 2.10, in four steps: A. Ants travelling between food and source (shortest path), B. An obstacle is placed, and the previous path is no longer available, C. Ants start going in different directions and pheromone levels are updated, D. The shortest distance is preferred as pheromone deposition has increased there. Furthermore, the mathematical analysis of the ACO are given as follows [35–37]:

The two important sub-problems that are required to be addressed here are:

1) The probability with which a particular route is chosen.
2) The rate at which the pheromone will evaporate.

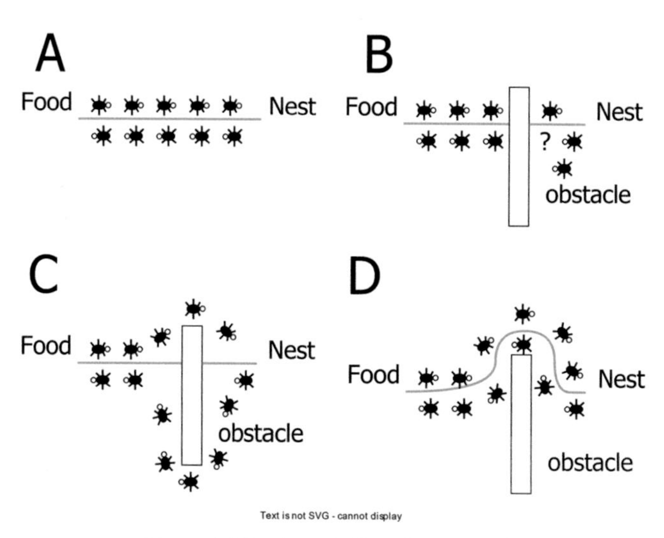

Figure 2.10 Ants finding shortest path.

In the ACO algorithms, the probability of an ant at node i to choose a route from node i to j is given by:

$$p_{ij} = \frac{\phi_{ij}^{\alpha} d_{ij}^{\beta}}{\sum_{i,j=1}^{n} \phi_{ij}^{\alpha} d_{ij}^{\beta}} \tag{2.16}$$

Here, $\alpha > 0$ and $\beta > 0$ are called the influence parameters. Typically, the values chosen for α and β are $\alpha \approx \beta \approx 2$. ϕ_{ij}^{α} represents the concentration of pheromone on the path connecting nodes i with node j, and d_{ij}^{β} represents the desirability of the route connecting node i with node j.

In addition, some known values for the route, for example, the distance is used so that $d_{ij}^{\beta} \propto 1/s_{ij}^{\beta}$. This represents that routes with lesser distances can be chosen with higher probability since the distance is smaller. Therefore, the strength of trail is stronger. Thus, this formula aptly reflects the condition that ants should follow the paths with stronger trail.

In case when $\alpha = \beta = 1$, the probability of choosing a path becomes directly proportional to the pheromone concentration. The concentration of pheromone decays with time which is due to the process of evaporation of pheromones. This saves the system from being trapped in local optima. In a situation where pheromone evaporation rate is 0, the first path that is randomly chosen by the first ant. It will then be used by all other ants irrespective of whether it is the shortest one.

For a rate of γ of constant evaporation, the concentration of pheromone mostly varies exponentially with respect to time, given by:

$$\phi(t) = \phi_0 e^{-\gamma t} \tag{2.17}$$

Here, ϕ_0 is the concentration of pheromone initially. If $\gamma t << 1$, then we have $\phi(t) \approx (1 - \gamma t)\phi_0$. For Δt = 1 (unit time increment), the evaporation rate is approximated by $\phi^{t+1} \leftarrow (1-\gamma)\phi^t$. Therefore, simplified pheromone update formula is given by:

$$\phi_{ij}^{t+1} = (1 - \gamma)\phi_{ij}^t + \delta\phi_{ij}^t, \tag{2.18}$$

where $\gamma \in [0, 1]$ specifies the pheromone evaporation rate.

The increment term $\delta\phi_{ij}^t$ gives us the amount of pheromone deposited at a particular time t along a route connecting node i with node j when an ant travels a distance L. Generally, $\delta\phi_{ij}^t \propto 1/L$. Therefore, the pheromone deposit is zero when no ants are on a route.

2.6.4 Grey wolf optimization

Grey wolf optimization (GWO) algorithm is a type of swarm intelligence (SI) that is based-on population-based metaheuristics methods. The GWO is a high-level procedure to find the solution based on the natural power hierarchy in the group of animals or natural predators [38]. It simulates the social dominance hierarchy along with hunting and attacking strategies of grey wolves. Like most wild hunting animals, the grey wolves also live in herds and have access to resources according to their position level in a hierarchy of social dominance. This hierarchy is set according to their strength, agility, and management abilities, as shown in Figure 2.11.

Here, the alpha wolf is the single most dominant and they are the only leader. Alpha makes important decisions for the pack like choosing the target and method of hunting, a safe place to rest, time to wake, and so on. The alpha is the dictator of the pack, but democratic behaviour is also present throughout the pack, where the alpha follows the suggestions of other wolves. The alpha is not the strongest member but is better at managing the pack. So, the core structure of the pack is organization and discipline instead of strength.

Beta wolves are at the second level of the dominance hierarchy. They are considered subordinates that help the alpha in decisions and other activities. Betas are the candidates for the position of alpha in their absence. Beta wolves

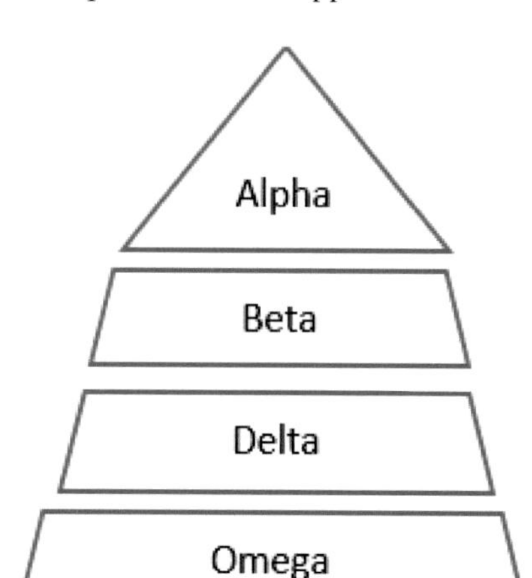

Figure 2.11 Social dominance hierarchy in a grey wolves' pack.

respect the alpha but command it from lower-level wolves. Betas can be considered messengers of alpha and gives feedback to the alpha.

Delta wolves can be considered scouts, caretakers, and hunters. Delta wolves must report to alpha and beta wolves, but they are still more dominant than the submissive wolves of the pack. They are the security guards of the pack; they watch the territory and the potential signs of danger. They are responsible for hunting.

The least dominant or the most submissive grey wolves are omegas. The omegas are used as a scapegoat. They are most likely to get sacrificed for the good of the pack. They are used as prey to attract other prey. They must report to all the wolves. It is assumed that omegas are redundant because their existence soaks the violence and frustration of the pack [7, 8, 38, 39]. The pseudo-code of the GWO is also given in code 2.1 on page 36.

The main phases of the hunting process of grey wolves are:

- Locating and surrounding the prey
- Harassing the prey
- Attacking the prey

The social hierarchy and mathematical modelling of the GWO are given as:

2.6.4.1 Social fitness hierarchy of optimal solutions:

- Optimally best – Alpha (α)
- Second best – Beta (β)

- Third best – Delta (δ)
- Least optimal – Omega (ω)

2.6.4.2 Approaching and surrounding the prey:

The mathematical modelling of the encircling prey during the hunt is given as:

$$\vec{A} = \left|\vec{B} \cdot \vec{Z}_p(t) - \vec{Z}_p(t)\right| \tag{2.19}$$

$$\vec{Z}(t+1) = \vec{Z}_p(t) - \vec{D} \cdot \vec{A} \tag{2.20}$$

$$\vec{D} = 2\vec{a} \cdot \vec{r}_1 - \vec{a} \text{ and } \vec{B} = 2\vec{r}_2 \tag{2.21}$$

where

t: Current iteration

$\vec{D}$ and $\vec{B}$: Coefficient vector

$\vec{Z}_p$: Position vector of prey

$\vec{Z}$: Position vector of grey wolf

2.6.4.3 Harassing prey

In every turn, omega wolves must change their positions according to the above hierarchical inhabitant's position because wolves with above hierarchy have more knowledge to adjust according to the prey.

$$\vec{A_\alpha} = \left|\vec{B}_1 \cdot \vec{Z_\alpha}(t) - \vec{X}(t)\right|, \vec{A_\beta} = \left|\vec{B}_2 \cdot \vec{Z_\beta}(t) - \vec{Z}(t)\right|,$$
$$\vec{A_\delta} = \left|\vec{B}_3 \vec{Z}_\delta(t) - \vec{Z}(t)\right| \tag{2.22}$$

$$\vec{Z}_1(t+1) = \vec{Z_\alpha}(t) - \vec{D}_1 \cdot \vec{A_\alpha}, \vec{Z}_2(t+1) = \vec{Z_\beta}(t) - \vec{D}_2 \cdot \vec{A_\beta},$$
$$\vec{X}_3(t+1) = \vec{Z_\delta}(t) - \vec{D}_3 \cdot \vec{A_\delta} \tag{2.23}$$

$$\vec{Z}(t+1) = \left(\vec{Z}_1 + \vec{Z}_2 + \vec{Z}_3\right)/3. \tag{2.24}$$

2.6.4.4 Attacking prey (exploitation)

Wolves start attacking the prey when it stops moving, so the value of a decreases. This change also alters the interval of D, i.e., when $\vec{a}$ decrease from 2 to 0, then the interval of $\vec{D}$ will limit to $[-a, \quad a]$. If $\vec{D}$ is in the interval of $[-1, 1]$ then the next position of the wolf will be the position of prey; it means the wolf attacks the prey.

2.6.4.5 Searching for prey (exploration)

Searching operation for locating the prey is done with respect to the position of grey wolves as per hierarchical level. For searching a new prey, grey wolves diverge from the prey. It can be represented in mathematical form the values of $\vec{D}$ will be greater than 1 and less than –1 in this divergence. The values of $\vec{B}$ will also limit to the interval of 0 to 2.

Pseudocode 2.1: Grey Wolf Optimization

1: Randomizing inhabitants in Grey wolves Z_i (i=1, 2, 3, 4, ……,n)
2: Initializing the coefficient vector a, $\vec{D}$ and $\vec{B}$.
3: Discerning individual grey's fitness in the pack
    Z_α=member with best fitness
    Z_β=second best-fit member
    Z_δ=third best fit member
4: FOR t = 1 to MaxNumberOfIterations:
    Changing the places of all search agent by Eq. (2.20).
    Updating coefficient Vectors, a, $\vec{D}$ and $\vec{B}$
    Work out Wellness of all specialists participating in search operation
    Change Z_α, Z_β, Z_δ with new Values.
  END FOR
5: return Z_α

2.7 Simulated Annealing

Simulated annealing (SA) algorithm solves unconstrained optimization problems with limited development and mimics the annealing process in the metallurgy to improve a global search. The annealing process was proposed by Metropolis et al. [40] in the metallurgy field. Some of the advantages of SA are found in literature such as: It is a global optimization technique, requires less memory, an iterative improvement algorithm etc. [41, 42]. The working principle of the SA algorithm is given in the following steps for minimizing the $y = H(x)$, subject to the constraint $x^{\min} < x < x^{\max}$.

Step 1: Let us start with initial high temperature of molten metal: T_o and consider randomly generated initial solutions x_o; Termination criterion: $\in$ (a small number); Set iteration number: t=0.

Step 2: Now, the algorithms calculate the temperature of $(t + 1)$th iteration which is $T_{t+1} = 0.5T_t$ as 50% of the previous iterations. This reduction in temperature can be different as per requirements. Based on the $(t+1)$th temperature, we can generate a solution i.e., $(t + 1)$th iteration (x_{t+1}) at random in the neighbourhood of X_t.

Step 3: In this step, the algorithm is comparing the energy with previous energy that if the change in the energy $dH = H(X_{t+1}) - H(X_t) <0$, then, algorithm accepts the solution X_{t+1} as the next solution and set $t = t + 1$. Otherwise, generate a random number (r) which is in the range of (0.0, 1.0). In addition, if $r <= e^{-\mathrm{dE}/T_{\mathrm{t}+1}}$, we accept X_{t+1} as the next solution and set $t = t + 1$, else, algorithm reject X_{t+1} and set $t = t$ and go to step 2.

Step 4: If $| H(X_{t+1}) - H(X_t)| < \in$ and T reaches a small value, we terminate the SA program.

2.8 Conclusion

In this chapter, a detailed discussion on different soft computing techniques is presented. These techniques can be applied to power system applications in different ways. The mathematical equations and advantages in these techniques are expressed and elaborated as much as possible. The fuzzy logic is mostly used for control applications along with clustering etc. In the neural network, the different neural networks structures are presented and generally BP is used to minimize the errors. Furthermore, nontraditional optimization techniques are also covered to overcome the exiting problems. Neuro-fuzzy i.e., ANFIS is preferred that takes advantage of both the concepts. This chapter gives the basics of soft computing techniques that can be directly utilized for power system applications.

Acknowledgements

The authors are thankful to the editors for providing an opportunity for this book.

References

[1] L. A. Zadeh, "Fuzzy sets," Information and Control, vol. 8. pp. 338–353, 1965, doi: 10.1109/2.53.

[2] L. A. Zadeh, "Soft computing and fuzzy logic," Fuzzy sets, fuzzy logic, and fuzzy systems: selected papers by Lotfi A Zadeh, World Scientific, 1996, pp. 796–804.

[3] A. Kumar and V. Kumar, "Evolving an interval type-2 fuzzy PID controller for the redundant robotic manipulator," Expert Syst. Appl., vol. 73, pp. 161–177, 2017, doi: 10.1016/j.eswa.2016.12.029.

[4] A. Kumar and V. Kumar, "Hybridized ABC-GA optimized fractional order fuzzy pre-compensated FOPID Control design for 2-DOF robot manipulator," AEUE - Int. J. Electron. Commun., vol. 79, pp. 219–233, 2017, doi: 10.1016/j.aeue.2017.06.008.

[5] O. Castillo, L. Cervantes, J. Soria, M. Sanchez, and J. R. Castro, "A generalized type-2 fuzzy granular approach with applications to aerospace," Inf. Sci. (Ny)., vol. 354, pp. 165–177, 2016, doi: 10.1016/j.ins.2016.03.001.

[6] C. J. C. Burges, "A tutorial on support vector machines for pattern recognition," Data Min. Knowl. Discov., vol. 2, no. 2, pp. 121–167, 1998.

[7] A. Kumar and R. Raj, "Design of a fractional order two layer fuzzy logic controller for drug delivery to regulate blood pressure," Biomed. Signal Process. Control, vol. 78, 2022.

[8] R. Sharma and A. Kumar, "Optimal Interval Type-2 Fuzzy Logic Control Based Closed-Loop Regulation of Mean Arterial Blood Pressure using the Controlled Drug Administration," IEEE Sens. J., vol. 22, no. 7, pp. 7195–7207, 2022, doi: 10.1109/JSEN.2022.3151831.

[9] F. Y. Xu, X. Cun, M. Yan, H. Yuan, Y. Wang, and L. L. Lai, "Power market load forecasting on neural network with beneficial correlated regularization," IEEE Trans. Ind. Informatics, vol. 14, no. 11, pp. 5050–5059, 2018.

[10] L. Zhang, G. Wang, and G. B. Giannakis, "Real-time power system state estimation and forecasting via deep unrolled neural networks," IEEE Trans. Signal Process., vol. 67, no. 15, pp. 4069–4077, 2019.

[11] S. Wen, Y. Wang, Y. Tang, Y. Xu, P. Li, and T. Zhao, "Real-time identification of power fluctuations based on LSTM recurrent neural network: A case study on Singapore power system," IEEE Trans. Ind. Informatics, vol. 15, no. 9, pp. 5266–5275, 2019.

[12] G. Bebis and M. Georgiopoulos, "Feed-forward neural networks," IEEE Potentials, vol. 13, no. 4, pp. 27–31, 1994.

[13] J. Barzilai, "On neural-network training algorithms," Human-Machine Shared Contexts, Academic Press, Elsevier, 2020, pp. 307–313.
[14] H. Wu, Y. Zhou, Q. Luo, and M. A. Basset, "Training feedforward neural networks using symbiotic organisms search algorithm," Comput. Intell. Neurosci., vol. 2016, 2016.
[15] A. Fadlalla and C.-H. Lin, "An analysis of the applications of neural networks in finance," Interfaces (Providence)., vol. 31, no. 4, pp. 112–122, 2001.
[16] A. A. Gunn, "The diagnosis of acute abdominal pain with computer analysis.," J. R. Coll. Surg. Edinb., vol. 21, no. 3, pp. 170–172, 1976.
[17] O. I. Abiodun, A. Jantan, A. E. Omolara, K. V. Dada, N. A. Mohamed, and H. Arshad, "State-of-the-art in artificial neural network applications: A survey," Heliyon, vol. 4, no. 11, p. e00938, 2018.
[18] W. De Mulder, S. Bethard, and M.-F. Moens, "A survey on the application of recurrent neural networks to statistical language modeling," Comput. Speech Lang., vol. 30, no. 1, pp. 61–98, 2015.
[19] K. Chen, Y. Zhou, and F. Dai, "A LSTM-based method for stock returns prediction: A case study of China stock market," in 2015 IEEE international conference on big data (big data), 2015, pp. 2823–2824.
[20] A. H. Fath, F. Madanifar, and M. Abbasi, "Implementation of multilayer perceptron (MLP) and radial basis function (RBF) neural networks to predict solution gas-oil ratio of crude oil systems," Petroleum, vol. 6, no. 1, pp. 80–91, 2020.
[21] D. Broomhead and D. Lowe, "Multivariable functional interpolation and adaptive networks, complex systems," Complex Syst., vol. 2, pp. 321-355., 1988.
[22] H. Faris, I. Aljarah, and S. Mirjalili, "Evolving radial basis function networks using moth–flame optimizer," Handbook of neural computation, Elsevier, 2017, pp. 537–550.
[23] P. Kumudha and R. Venkatesan, "Cost-sensitive radial basis function neural network classifier for software defect prediction," Sci. World J., vol. 2016, 2016.
[24] J.-S. R. Jang, "ANFIS: adaptive-network-based fuzzy inference system," IEEE Trans. Syst. Man. Cybern., vol. 23, no. 3, pp. 665–685, 1993.
[25] J.-S. R. Jang, "Self-learning fuzzy controllers based on temporal back-propagation," IEEE Trans. Neural Networks, vol. 3, no. 5, pp. 714–723, 1992.

[26] D. Karaboga and E. Kaya, "Adaptive network based fuzzy inference system (ANFIS) training approaches: a comprehensive survey," Artif. Intell. Rev., vol. 52, no. 4, pp. 2263–2293, 2019.
[27] Y. Yan, L. Wang, T. Wang, X. Wang, Y. Hu, and Q. Duan, "Application of soft computing techniques to multiphase flow measurement: A review," Flow Meas. Instrum., vol. 60, pp. 30–43, 2018.
[28] J. Saroha, M. Singh, and D. K. Jain, "A New Adaptive Neuro-Fuzzy Inference System (ANFIS) Controller to Control the Power System equipped by Wind Turbine," IEEE Trans. Ind. Informatics, vol. 14, no. 12, pp. 5338–5345, 2018.
[29] C. Cortes and V. Vapnik, "Support-vector networks," Mach. Learn., vol. 20, no. 3, pp. 273–297, 1995.
[30] Y. Tian, Y. Shi, and X. Liu, "Recent advances on support vector machines research," Technol. Econ. Dev. Econ., vol. 18, no. 1, pp. 5–33, 2012.
[31] J. H. Holland, "Adaptation in Natural and Artificial Systems. ann arbor: University of Michigan Press," University of Michigan Press, 1975.
[32] D. E. Golberg, Genetic algorithms in search, optimization, and machine learning. 1989.
[33] R. Eberhart and J. Kennedy, "Particle swarm optimization," in Proceedings of the IEEE International Conference on Neural Networks, 1995, vol. 4, pp. 1942–1948.
[34] M. Dorigo, "Optimization, learning and natural algorithms," 1992.
[35] M. Dorigo, V. Maniezzo, and A. Colorni, "Ant system: optimization by a colony of cooperating agents," IEEE Trans. Syst. Man, Cybern. Part B, vol. 26, no. 1, pp. 29–41, 1996.
[36] M. Dorigo, G. Di Caro, and L. M. Gambardella, "Ant algorithms for discrete optimization," Artif. Life, vol. 5, no. 2, pp. 137–172, 1999.
[37] C. Blum, "Ant colony optimization: Introduction and recent trends," Phys. Life Rev., vol. 2, no. 4, pp. 353–373, 2005.
[38] S. Mirjalili, M. S. Mirjalili, and A. Lewis, "Grey Wolf Optimizer," Adv. Eng. Softw., vol. 69, pp. 46–61, 2014, doi:10.1016/j.advengsoft.2013.12.007.
[39] R. E. Precup, R. C. David, and E. M. Petriu, "Grey Wolf optimizer algorithm-based tuning of fuzzy control systems with reduced parametric sensitivity," IEEE Trans. Ind. Electron., vol. 64, no. 1, pp. 527–534, 2017, doi: 10.1109/TIE.2016.2607698.

[40] N. Metropolis, A. W. Rosenbluth, M. N. Rosenbluth, A. H. Teller, and E. Teller, "Equation of state calculations by fast computing machines," J. Chem. Phys., vol. 21, no. 6, pp. 1087–1092, 1953.
[41] D. Delahaye, S. Chaimatanan, and M. Mongeau, "Simulated annealing: From basics to applications," Handbook of Metaheuristics, Springer, 2019, pp. 1–35.
[42] C. Venkateswaran, M. Ramachandran, K. Ramu, V. Prasanth, and G. Mathivanan, "Application of Simulated Annealing in Various Field," Mater. its Charact., vol. 1, no. 1, pp. 1–8, 2022.

3

Load Flow Solution Algorithm for AC-DC Radial Distribution Systems in the Presence of Distributed Generation

Krishna Murari[1], Sameep Sahu[2], Om Hari Gupta[2], N.P. Padhy[3], S. Kamalasadan[4]

[1]Electrical Engineering and Computer Science Department, University of Toledo, Ohio, United States of America
[2]Department of Electrical Engineering, NIT Jamshedpur, India
[3]Department of Electrical Engineering, Indian Institute of Technology Roorkee, India
[4]Department of Electrical and Computer Engineering, University of North Carolina at Charlotte, United States of America
Email: krishna.murari@utoledo.edu1, 2021pgeeps15@nitjsr.ac.in, omhari.ee@nitjsr.ac.in, nppadhy@ee.iitr.ac.in, skamalas@charlotte.edu

Abstract

A load flow (LF) methodology for an AC-DC radial distribution system (DS) is indicated in this chapter. A p.u. model of power system converters is introduced to bridge the difference between an AC and DC DS to carry out the LF of the DS. The different ways that power converters can operate has also been considered to show the effectiveness of the indicated approach. The sensitivity and breakpoint matrix have been designed to access the LF solution of DS with PV and V^{dc} bus. The indicated method is computationally effective due to the concise algebraic operation and effective search technique.

Keywords: Load flow, AC-DC radial distribution system, power system converters

3.1 Introduction

Over the past few decades, there has been a competition between AC and DC distribution systems. While AC systems were preferred over DC systems decades ago, the latter has recently regained popularity in certain applications such as long-distance power transmission and storage systems, as DC loads have increased due to the incorporation of renewable energy technologies [1–4]. However, to use DC equipment, the available AC power must be rectified into DC power, which requires energy conversion stages using rectifiers. Moreover, the DC power from renewable units needs to be converted to AC before it can be connected to the existing power grid and then converted back to DC for end-users, which can result in significant energy losses and synchronization problems. As a result, DC grids are becoming more attractive. However, despite the many advantages of the DC system, the historical importance and impact of the AC system cannot be ignored. Therefore, researchers are now focused on designing AC-DC distribution systems that incorporate both AC and DC components.

A load flow (LF) study must be conducted for the optimal planning, assessment, and operation of a transmission and distribution system. The available LF methods for transmission systems [5–9] do not apply to a DS because of its greater *R/X* ratio and primarily radial architecture.

Due to the requirement of synchronization of grids fed through power converters, the LF analysis for a distribution system is significantly more complicated than for a pure AC DS. The proper converter is required to ensure that the algorithm is both reliable and robust. There exists a large volume of literature for LF solutions of AC DS but for AC-DC DS, only a few surveys have been done. The LF techniques implemented and mentioned in the literature review fall into three broad categories viz. Newton–Raphson and fast-decoupled [10–17], backward-forward sweep [18–21], and direct LF [22] based method. However, due to the inclusion of power converters and other uncertainties, the general LF methods cannot be applied to AC-DC DS but the method employing the backward/forward sweep method can be modified to yield the desired solution.

The term "backward-forward swap method" is deliberately selected for a reason. In the backward sweep, we derive the branch current starting from the furthest sections away from the slack bus in the AC-DC distribution system. Similarly, in "forward sweep" process we establish the unique equation used to calculate new bus voltages, starting from the slack bus area of the AC-DC distribution system. Notably, the presented method utilizes only elementary algebraic matrix operations and multiple graphical search procedures to attain the preferred LF solution for DS. This study aims to establish a method that directly solves the LF problem by integrating the dynamical aspects of distribution systems. As a result, the suggested approach avoids the LU-decomposition and backward/forward replacement of the admittance matrix or Jacobian matrix., as intended by conventional NR algorithms. The primary input required is the standard bus-branch-centred data. The load beyond branch matrix (LB), Load current matrix (LC), path impedance matrix (PI), Path drop matrix (PD), and Load flow matrix (LFM) have been utilized to compute the AC-DC system steady state parameters. This task can incorporate multiple power converters and their working modes. A modified breakpoint injection matrix was established for the power flow employing controlled distributed generation (DG). The effective operation makes the technique computationally efficient. The outcomes highlight the accuracy of the suggested algorithm.

3.2 LF Algorithm for AC-DC Radial Distribution Systems

A LF analysis is necessary to evaluate the functional states of such a system. To assess the system state of DS, it is crucial to solve the LF equations of the DS. Figure 3.1 depicts a one-line diagram of the DS to illustrate the developed LF algorithm. The architecture of AC-DC DS differs from that of AC DS. It comprises numerous AC and DC components, such as generating units, lines, loads, and buses. The DN is divided into several sub-regions for our purposes.

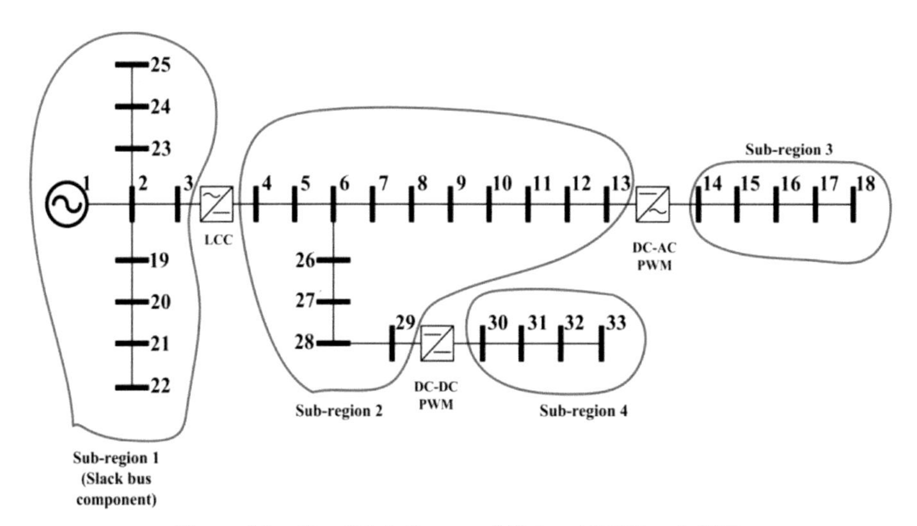

Figure 3.1 Simplified diagram of 33-bus AC-DC radial DS.

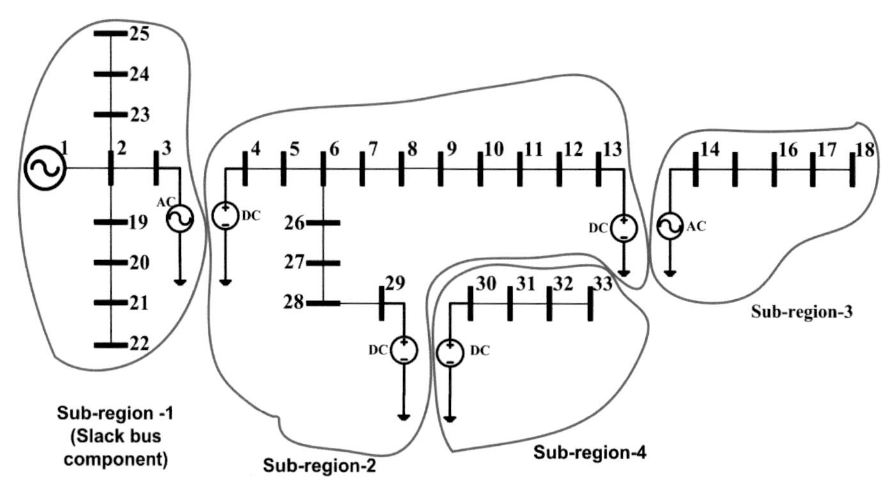

Figure 3.2 Simplified diagram of a radial DS.

The DS comprises four sub-regions that can be grouped into AC and DC bus-sections as demonstrated in Figure 3.2. These sub-DS are capable of being interconnected in a wide range of hybrid layouts. In the DS, the accurate design of converters will simplify LF equations. This section proposes p.u. equivalent models of three-phase PWM AC-DC converters, PWM DC-DC converters, and three-phase AC-DC bridge LCC converters to reduce the LF problem in AC-DC radial DS.

3.2.1 Converters modelling

The mathematical modelling of various types of converters is briefly discussed in this section. It involves the calculation of AC-DC LF under a balanced network environment.

3.2.1.1 Modelling of three-phase AC-DC bridge converters

The three-phase bridge converter can be considered as an AC-DC transformer. The effective turns ratio varies as the amount of current passing through the converter changes. Therefore, the AC-DC bridge converter has nonlinear transformation characteristics. Figure 3.3 depicts the equivalent circuit p.u. of a three-phase bridge converter (b).

The pictorial representation portrayed in Figure 3.3 (b) is in per-unit and quantities shown in Figure 3.3 (a) are phasor values. The equivalent mathematical model of a three-phase bridge converter is expressed as:

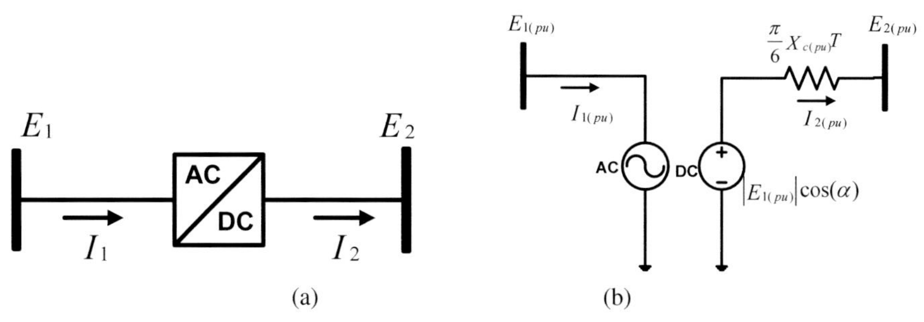

Figure 3.3 Three–phase bridge converter: (a) Simplified diagram, (b) p.u. equivalent diagram.

$$E_{2(pu)} = \left|E_{1(pu)}\right| \cos(\alpha) - \frac{\pi}{6} X_{c(pu)} I_{2(pu)}\, T. \tag{3.1}$$

By equating power on each side of a full bridge rectifier, the following equation for evaluating the behaviour of a bridge rectifier is obtained.

$$\left|E_{1(pu)}\right| \left|I_{1(pu)}\right| \cos(\phi)\eta = E_{2(pu)} I_{2(pu)}, \tag{3.2}$$

where
η = Converter's efficiency
E_1 = AC-side phasor voltage
E_2 = DC-side voltage
I_1 = Input current to the converter
I_2 = Output current from the converter
X_c = Commutation reactance
α = Angle of commutation
T = 1 (rectification mode) and –1 (inverting mode).

The following equation expresses the linkage between I_1 and I_2

$$\left|I_{1(pu)}\right| = \frac{I_{2(pu)}}{\eta}, \tag{3.3}$$

$$\cos(\alpha) \approx \cos(\phi). \tag{3.4}$$

Four control strategies can be deployed in these converters i.e., fixed DC voltage, current, power, and minimum firing angle.

The base equations are:

$$P_{\mathrm{dc(B)}} = VA_{\mathrm{ac(B)}} \tag{3.5}$$

$$I_{\mathrm{dc(B)}} = \frac{\pi}{\sqrt{6}} I_{\mathrm{ac(B)}} \tag{3.6}$$

$$V_{\mathrm{dc(B)}} = \frac{3\sqrt{2}}{\pi} V_{\mathrm{ac(B)}} \tag{3.7}$$

$$Z_{\mathrm{dc(B)}} = \frac{V_{\mathrm{dc(B)}}}{I_{\mathrm{dc(B)}}} = \frac{18}{\pi^2} Z_{\mathrm{ac(B)}}. \tag{3.8}$$

Here, $VA_{\mathrm{ac(B)}}$, $V_{\mathrm{ac(B)}}$, $I_{\mathrm{ac(B)}}$, and $Z_{\mathrm{ac(B)}}$ portray the three-phase base power, voltage, current, and impedance respectively of the converter on the AC-side and $P_{\mathrm{dc(B)}}, V_{\mathrm{dc(B)}}, I_{\mathrm{dc(B)}}$ and $Z_{\mathrm{dc(B)}}$ denotes the base power, voltage, current, and impedance respectively of the converter on DC-side.

3.2.1.2 Modelling of PWM AC-DC converter

Figures 3.4 (a) and (b) show a pictorial representation of the PWM converter and its p.u. equivalent diagram. The notations/quantities depicted in Figure 3.4 (b) are in p.u. and quantities shown in Figure 3.4 (a) are phasor values. The p.u. equivalent model (Figure 3.4 (b)) of a three-phase PWM converter can be expressed as:

$$\left|E_{2(pu)}\right| = \left|E_{1(pu)}\right| (MI). \tag{3.9}$$

By equalizing the active power on each side of a PWM converter, the behaviour can be assessed.

$$\left|E_{2(pu)}\right| \left|I_{2(pu)}\right| \cos(\phi) = E_{1(pu)} I_{1(pu)} \eta, \tag{3.10}$$

where ∅ = angle between phase voltage and phase current.

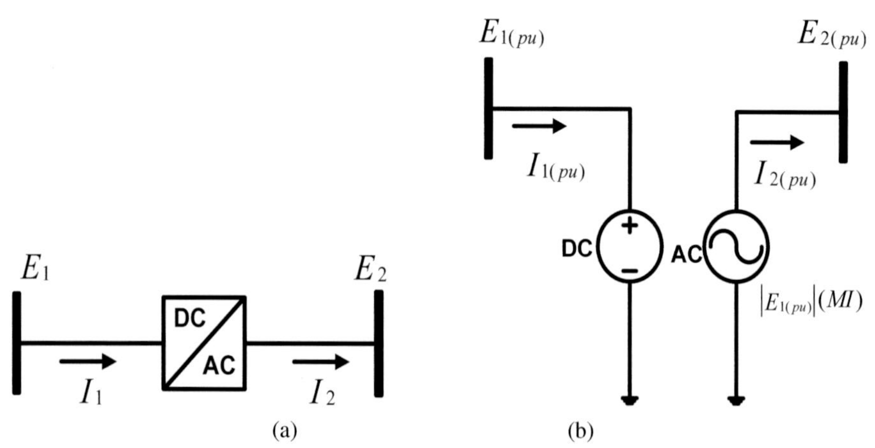

Figure 3.4 Three-phase PWM converter: (a) Schematic diagram, (b) p.u. equivalent diagram.

Substituting eqn (3.9) into eqn (3.10) generates an equation that describes the relationship between input current (I_1) and output current (I_2).

$$I_{1(pu)} = \frac{\left|I_{2(pu)}\right| \cos(\phi)}{\eta}(MI), \tag{3.11}$$

where
E_1 = DC-side voltage
E_2 = AC-side phasor voltage
I_1 = Input DC current to the converter
I_2 = Output AC phasor current from the converter (fundamental component)
MI = Modulation index
η = Efficiency of converter.

The following are the base values for a three-phase PWM converter described below:

$$P_{\mathrm{dc(B)}} = VA_{\mathrm{ac(B)}}, \tag{3.12}$$

$$I_{\mathrm{ac(B)}} = \frac{2\sqrt{2}}{3} I_{\mathrm{dc(B)}}, \tag{3.13}$$

$$V_{\mathrm{ac(B)}} = \frac{\sqrt{3}}{2\sqrt{2}} V_{\mathrm{dc(B)}}, \tag{3.14}$$

$$Z_{\mathrm{ac(B)}} = \frac{V_{\mathrm{ac(B)}}}{I_{\mathrm{ac(B)}}} = \frac{3}{8} Z_{\mathrm{dc(B)}}. \tag{3.15}$$

3.2.1.3 Modelling of PWM DC-DC converter

DC-DC converter [18] is used to regulate voltage. However, alternative control strategies, such as those using fixed power, current, and duty ratio (D), are also appropriate. The implementation of these converters has been discussed in continuous conduction mode. Figures 3.5(a) and (b) show a pictorial representation of the DC-DC converter and its p.u. equivalent diagram.

Buck, boost, and buck–boost converters (BBC) were designed utilizing the following mathematical equations:

$$\left|E_{2(pu)}\right| = \left|E_{1(pu)}\right|(A). \tag{3.16}$$

Another equation that helps evaluate the behaviour of the converter is attained by equalizing the power on each side.

$$E_{1(pu)} I_{1(pu)} \eta = E_{2(pu)} I_{2(pu)}. \tag{3.17}$$

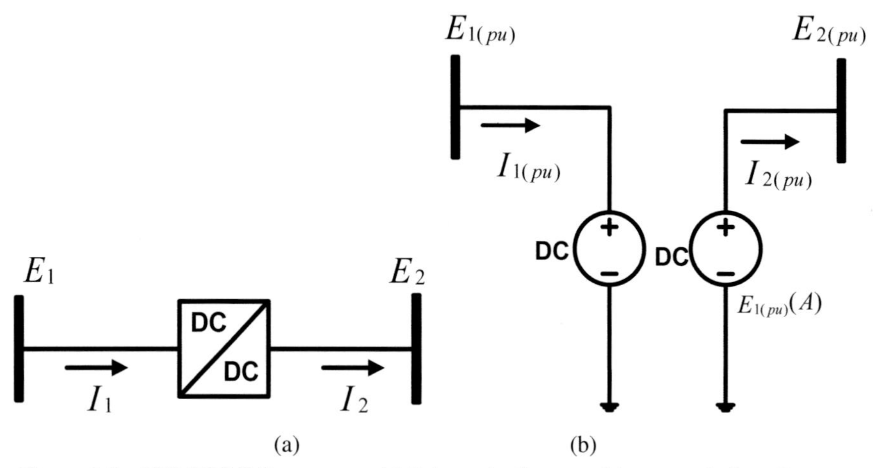

Figure 3.5 PWM DC-DC converter: (a) Schematic diagram, (b) p.u. equivalent diagram.

On substituting eqn (3.16) in eqn (3.17), the net equation reflects the relationship between input (I_1) and output (I_2) current.

$$I_{1(pu)} = A \times \frac{I_{2(pu)}}{\eta}, \tag{3.18}$$

where A is D, $\frac{1}{1-D}$, and $\frac{D}{1-D}$ for the buck, boost, and buck-boost operations respectively.

Here,

E_1 = Converter input voltage
E_2 = Converter output voltage
I_1 = DC current input
I_2 = DC current output
D = Duty ratio
η = Converter efficiency.

The detailed equations (phase and per-unit equations) of all kinds of converters are mentioned in Table 3.1.

Note: PWM and DC-DC converters are more efficient than bridge converters at the distribution level, allowing them to effectively transfer power between buses and the interface with distributed generation (DG) while meeting necessary control requirements. However, in cases where control considerations are not as critical, a bridge converter may be a suitable option for serving as a bus-to-bus interface between buses.

Table 3.1 Converters classification.

Equation types	Phase equation	Per-unit equation
Three-phase rectifier (LCC)		
Voltage	$E_2 = \frac{3\sqrt{6}}{\pi} \left\|E_{1(ph)}\right\| \cos(\alpha)$	$E_{2(pu)} = \left\|E_{1(pu)}\right\| \cos(\alpha)$
Current	$\|I_1\| = \frac{\sqrt{6}}{\eta\pi} I_2$	$\left\|I_{1(pu)}\right\| = \frac{1}{\eta} I_{2(pu)}$
Power balance	$3 \left\|E_{1(ph)}\right\| \|I_1\| \cos(\emptyset)\eta = (E_2 I_2)$	$\left\|E_{1(jw)}\right\| I_{1(jw)} \mid \cos(\phi)\eta = E_{2(jw)} I_{2(\rho w)}$
PWM DC-AC converter		
Voltage	$\left\|E_{2(ph)}\right\| = \frac{1}{2\sqrt{2}}(MI)E_1$	$\left\|E_{2(pu)}\right\| = E_{1(pu)}(MI)$
Current	$I_1 = \frac{3}{\eta 2\sqrt{2}}(M) \|I_2\| \cos(\emptyset)$	$I_{1(\mu')} = \frac{\left\|I_{2(\rho m)}\right\| \cos(\phi)}{\eta}(MI)$
Power balance	$3 \left\|E_{2(ph)}\right\| \|I_2\| \cos(\varnothing) = (E_1 I_1 \eta)$	$\left\|E_{2(\rho m)}\right\| I_{2(\rho j)} \mid \cos(\phi) = E_{1(ju)} I_{1(ju)} \eta$
PWM DC-DC converter		
Voltage	$\|E_2\| = \|E_1\| (A)$	$\left\|E_{2(\rho w)}\right\| = \left\|E_{1(\rho w)}\right\| (A)$
Current	$I_1 = A \times \frac{I_2}{\eta}$	$I_{1(p,u.)} = A \times \frac{I_{2(p)}}{\eta}$
Power balance	$(E_1 I_1 \eta) = (E_2 I_2)$	$E_{1(j\ \mathrm{m})} I_{1(\mathrm{pa})} \eta = E_{2(\mathrm{pr})} I_{2(\mathrm{pa})}$

3.2.2 Solution methodology

The suggested LF technique has been illustrated using the one-line diagram in Figure 3.6. The AC-DC DS has been grouped into many subsystems. Each subsystem will act as an independent *D*. The output bus of the converter (bus 3d in Figure 3.6) will be considered as the slack bus for the sub-region. The formulation of the relevant matrices, as discussed earlier, must be performed in each sub-region separately.

Note **A**: The subscript "g" in matrices indicates the number of sub-regions. All significant matrices must be framed for each sub-region of DS.

Note **B**: It must be noted that if a specific symbol does not have a definition anywhere in the manuscript, it is regarded as a load flow calculation variable.

The implementation procedure for the suggested LF algorithm is summarized below:

Step 1. Determine the number of sub-regions within an AC-DC DS. In Figure 3.6, the DS comprises of two sub-regions.

Step 2. Specify all branches and nodes within the DS that correspond to their respective sub-region. Organize the load's past branch matrix (LB_g) separately for each sub-region of the AC-DC DS.

g = sub-region number = 1, 2, 3, …, C.

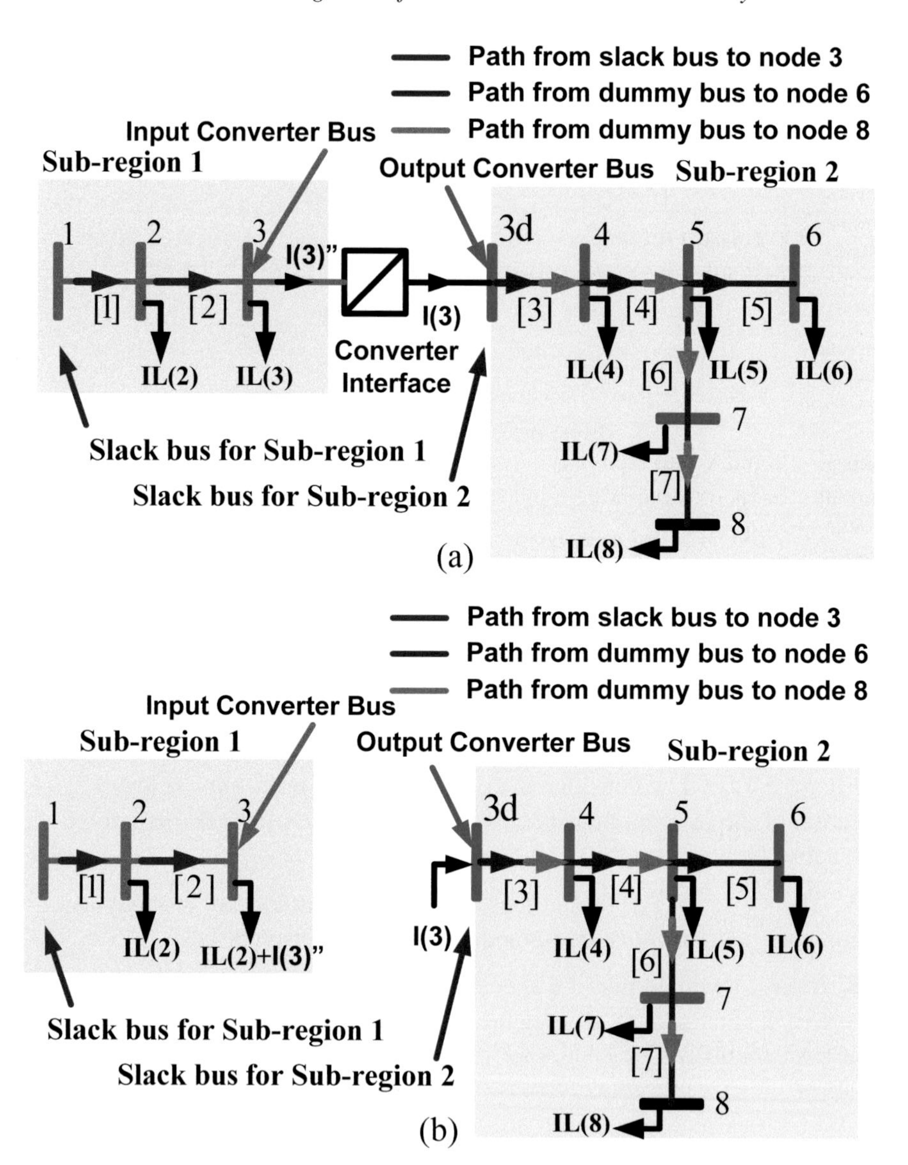

Figure 3.6 Pictorial representation of an AC-DC radial DS.

C = Total sub-regions

The LB_g matrices for the sub-region 1 and sub-region 2 are denoted as:

$$LB_1 = \begin{bmatrix} S(2) & S(3) \\ 0 & S(3) \end{bmatrix}, \tag{3.19}$$

$$LB_2 = \begin{bmatrix} S(4) & S(5) & S(6) & S(7) & S(8) \\ 0 & S(5) & S(6) & S(7) & S(8) \\ 0 & 0 & S(6) & 0 & 0 \\ 0 & 0 & 0 & S(7) & S(8) \\ 0 & 0 & 0 & 0 & S(8) \end{bmatrix}. \tag{3.20}$$

Step 3. For every sub-region, create the path-impedance matrix PI_g. The PI_g matrices of sub-regions 1 and 2 in Figure 3.6 are prescribed as follows:

$$PI_1 = \begin{bmatrix} Z(1) & Z(2) \end{bmatrix}, \tag{3.21}$$

$$PI_2 = \begin{bmatrix} Z(3) & Z(4) & Z(5) & 0 & 0 \\ Z(3) & Z(4) & 0 & Z(6) & Z(7) \end{bmatrix}. \tag{3.22}$$

Establish a rated voltage profile at each node (comprising nodes that serve as interfaces for converters), and then develop voltage matrices for each DS sub-regions. The order of the voltage matrix V_g is $(1 \times (n - 1))$. The tolerance limit is set to $\varepsilon \leq 0.0001$.

$n =$ Resultant buses in a sub-region g of the DS.

Voltage matrices for both sub-rcgions of the DS depicted in Figure 3.6 are as follows:

$$V_1 = \begin{bmatrix} V(2) & V(3) \end{bmatrix}, \tag{3.23}$$

$$V_2 = \begin{bmatrix} V(4) & V(5) & V(6) & V(7) & V(8) \end{bmatrix}. \tag{3.24}$$

Step 5. The iteration number is set at t = 1 [with t(max) = 100].

Determine the regulating parameters of each converter. For a three-phase bridge converter depicted in Figure 3.6, α is computed.

$$\alpha = \cos^{1} \left(\frac{V(3d)}{|V(3)|} \right). \tag{3.25}$$

Compute D, if the converter in Figure 3.6 is a DC-DC converter.

$$\begin{aligned} D &= \left(\tfrac{V(3d)}{V(3)} \right) \text{ (buck)} \\ D &= 1 - \left(\tfrac{V(3)}{V(3d)} \right) \text{ (boost)} \\ \mathrm{D} &= \left(\tfrac{V(3d)}{V(3)+V(3d)} \right) \text{ (buck} - \text{boost)} \end{aligned}. \tag{3.26}$$

Compute MI, if the converter in Figure 3.6 is a three-phase VSC-converter.

$$MI = \left(\frac{|V(3d)|}{V(3)} \right). \tag{3.27}$$

Step 6. Determine the load and feeder current matrix i.e., (LC_g) and (FC_g) for every sub-region. The LC_g and FC_g of the sub-region farthest away from the sub-region containing the slack bus will be analysed first. The process for calculating the LC_g and FC_g is depicted below:

$W_g = \text{Conjugate } (LB_g)$[for AC sub-regions] = LB_g[for DC sub-regions)
$V1_g = \text{Conjugate } (V_g)$[for AC sub-regions] = V_g[for DC sub-regions)

for i = 1,2..............*m.*

$$\left.\begin{aligned} LC_g(i,j) &= \frac{W_g(i,j)}{V1_g(i,j)} \text{ (if AC sub} - \text{region)} \\ LC_g(i,j) &= \frac{W_g(i,j)}{V1_g(i,j)} \text{ (if DC sub} - \text{region)} \end{aligned}\right\}. \tag{3.28}$$

Compute LC_g for all values of *n* and *m*.

$$j = 1, 2, \ldots, n\text{-}1.$$

Each element in the LC_g denotes load current. The summation of all load currents in a single row yields the feeder current for the branch.

$$FC_g(i) = \sum_{l=1}^{n-1} LC_g(i,l) \tag{3.29}$$

$$i = 1, 2, ..., m,$$

where
m = Branches in a sub-region g of the DN
n = Buses in a sub-region g of the DN

The LC_g and FC_g matrix of sub-region 2 are estimated first for the DN as depicted in Figure 3.6. The obtained results are:

$$LC_2 = \begin{bmatrix} IL(4) & IL(5) & IL(6) & IL(7) & IL(8) \\ 0 & IL(5) & IL(6) & IL(7) & IL(8) \\ 0 & 0 & IL(6) & 0 & 0 \\ 0 & 0 & 0 & IL(7) & IL(8) \\ 0 & 0 & 0 & 0 & IL(8) \end{bmatrix}, \tag{3.30}$$

$$FC_2 = \begin{bmatrix} I(3) \\ I(4) \\ I(5) \\ I(6) \\ I(7) \end{bmatrix}. \tag{3.31}$$

Step 7. Determine the input current to the bus-to-bus linking converter linking two separate sub-regions. In Figure 3.6, the converter input current is estimated using the equations described below:

$$|I(3")| = \frac{1}{\eta} I(3) (\text{For a three-phase bridge-converter}). \tag{3.32}$$

$$I(3") = \frac{(MI)}{\eta} |I(3)| \cos(\emptyset) \ (\text{For a three-phase VSC-converter}). \tag{3.33}$$

$$I(3") = \frac{A \times I(3)}{\eta} (\text{For a DC-DC converter}). \tag{3.34}$$

Here, A is D, $\frac{1}{1-D}$, and $\frac{D}{1-D}$ for the buck, boost, and buck−boost operations respectively.

Step 8. Assess the LC_g and FC_g matrix for each sub-region of the AC-DC DS. For sub-region 1 of DS depicted in Figure 3.6, the LC_g and FC_g matrices are determined. The derived results are as follows:

$$LC_1 = \begin{bmatrix} IL(2) & IL(3) + I(3^{"}) \\ 0 & IL(3) + I(3^{"}) \end{bmatrix}, \tag{3.35}$$

$$FC_1 = \begin{bmatrix} I(1) \\ I(2) \end{bmatrix}. \tag{3.36}$$

Step 9. Evaluate the path drop matrix (PD_g) for each sub-distribution network in the following step. In Figure 3.6, the PD_g matrices for the two sub-regions of the DN are shown as follows:

$$PD_1 = \begin{bmatrix} Z(1) \times I(1) & Z(2) \times I(2) \end{bmatrix}, \tag{3.37}$$

$$PD_2 = \begin{bmatrix} Z(3) \times I(3) & Z(4) \times I(4) & Z(5) \times I(5) & 0 \times I(6) & 0 \times I(7) \\ Z(3) \times I(3) & Z(4) \times I(4) & 0 \times I(5) & Z(6) \times I(6) & Z(7) \times I(7) \end{bmatrix}. \tag{3.38}$$

Step 10. The $SBOBD_g$ matrix for each sub-region needs to be framed. For example, $SBOBD_1$ for sub-region 1 of the DN in Figure 3.6 is given as:

$$SBOBD_1 = \begin{bmatrix} D(1) & D(1) + D(2) \end{bmatrix},\tag{3.39}$$

where $D(1) = I(1) \times Z(1),\ \ D(2) = I(2) \times Z(2)$.

Correspondingly, the $SBOBD_g$ for the remaining sub-regions are estimated.

Step 11. Using the procedure outlined below, compute the bus-voltage matrix for the AC sub-region linked to a slack bus.

$$LFM_g = (\mathrm{voltage at the slack bus} \times \mathrm{U_g}) - \mathrm{SBOBD_g},\tag{3.40}$$

U_g = Binary matrix having the same order as $SBOBD_g$.
$U_g(i, j) = 0$, if $PI_g(i, j)$ = is 0.
$U_g(i, j) = 1$, else.

The largest value is picked from the matrix T_g, and then responses are saved in the variable $V_g(\mathrm{new})$, which will ultimately result in the creation of a voltage matrix.

The $V_1(\mathrm{new})$ for the sub-region 1 of the DN in Figure 3.6 is denoted as:

$$V_{1(new)} = \begin{bmatrix} V(2) & V(3) \end{bmatrix}.\tag{3.41}$$

Step 12. The output voltage is calculated for AC-DC converters linking two sub-regions. The regulating variables of all bus-to-bus linking converters must be determined.

Q is to be held constant at the output terminal of the converter for fixed voltage, then $3d$:

$$|V(3d)| = Q$$

$V(3)$ = Computed voltage at bus-3 {See (4.41)}.

Then,

Estimate α by eqn (3.25), for a three-phase bridge converter illustrated in Figure 3.6

Estimate MI by eqn (3.27), for a three-phase VSC converter illustrated in Figure 3.6

Estimate D by eqn (3.26), for a DC-DC converter illustrated in Figure 3.6.

The control limits of computed regulating parameters are examined in the subsequent phase. Suppose the controlled variable exceeds the limit, it will be fixed to that limit, signifying that this control scheme cannot be executed (Figure 3.7).

Once the governing parameters are obtained, it is simple to compute the output converter voltage utilizing the corresponding converter voltage equation (see eqn (3.1), (3.9), and (3.16)). The converter output voltage is estimated using the power factor angle difference at the converter terminal calculated at the beginning of iteration and in the current scenario. The difference is the new power factor angle output voltage of the converter.

Step 13. The output buses of the converter are classified as slack buses for the sub-region to which they belong. Consequently, the matrices U_g of all sub-regions connected with DN must be multiplied against corresponding converter output bus voltages.

Step 14. Refer to step 12 for the calculation of the revised bus-voltage matrix for the remaining sub-regions.

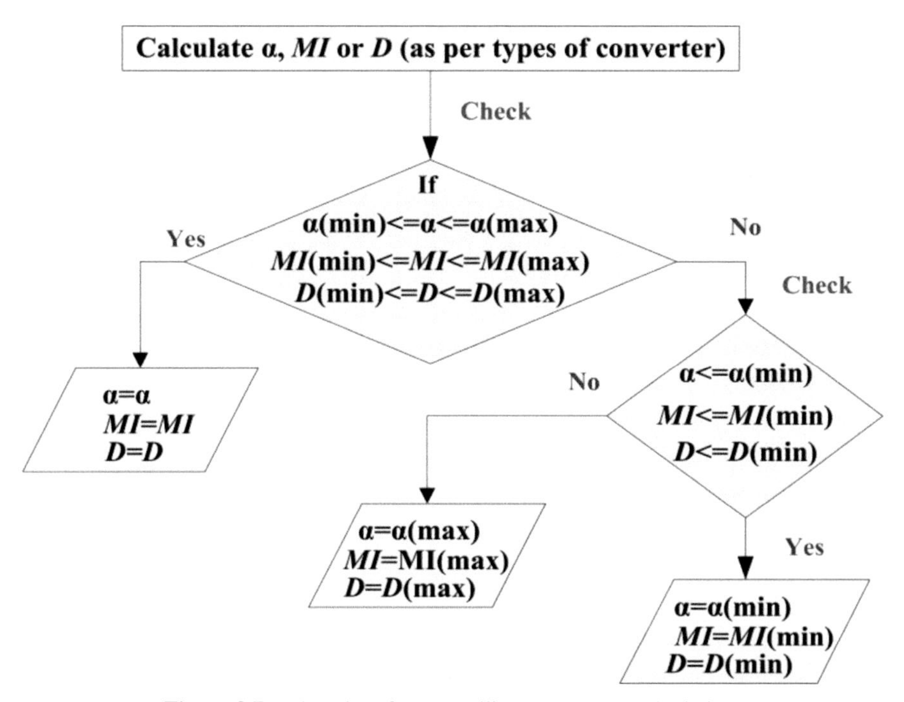

Figure 3.7 Flowchart for controlling parameters calculation.

Step 15. Examine the convergence criterion for each sub-region.

Step 16. *DVMAX* is estimated by taking the maximum value of convergence of all sub-regions.

Step 17. If $DVMAX < \varepsilon$ move to step 22, otherwise go to step 19.

Step 18. Substitute each sub-regions existing voltage matrix with its newly acquired voltage matrix.

$$\text{i.e.,} V_g = V_{g(\text{new})} \\ g = 1,\ 2C. \tag{3.42}$$

Step 19. $t = t\ +\ 1.$

Step 20. If $t^2 t(\text{maximum})$ move to step 6 else move to step 22.

Step 21. The output is displayed as "Solution converged."

Step 22. The output is displayed as "Solution not converged."

The algorithm is based on the p.u. equivalent model of numerous types of converters mentioned in this chapter. The p.u. converter equations utilized in Section 3.2 can be replaced with their phase equivalent equations mentioned in Table 3.1 and LF iteration will be executed accordingly.

3.3 LF Algorithm for AC-DC Radial Distribution System with DGs

The operation of distributed generation (DG) depends on the control strategy employed. For small DG units, they can be configured to deliver power at a consistent power factor, while larger DG units can be set to maintain a constant voltage. Depending on the control strategy chosen, the DG units connected to the AC bus can be represented as PQ bus/load bus or as PV bus. At PQ buses, the real and reactive power values are given, and the computation task involves determining the voltage magnitude and its phase angle. Meanwhile, at PV buses, the parameters to be determined are the reactive power and the voltage phase angle. Similarly, the DGs linked to the

DC bus can be modelled as either constant real power (P) loads or as P buses injecting real power into the node, along with voltage control V^{dc} nodes or buses. At P buses, the real power values are given, and the computation task involves determining the voltage magnitude. Meanwhile, at V^{dc} buses, the parameters to be determined are the real power and the voltage magnitude. The solution of the power flow problem in the presence of these DGs or DERs models will be explored in the following subsections. The stated method employs PQ, P, Vdc, and PV bus models of DG to simulate the injection of DGs into distribution networks. The DG developed as PQ and P buses may be integrated into the suggested LF algorithm by viewing the DGs' injection as a negative load. The power injected by DGs must be interpreted in the distribution network LB matrix. The right quantity of power injection must mitigate the actual and desired voltage difference when V^{dc} nodes are in DC region.

3.3.1 Case 1: The AC-DC distribution system with one V^{dc} node

Consider the DS represented in Figure 3.8, where a DG modelled as $\mathbf{V}^{dc}$ is supplying power at bus 8.

The desired quantity of real power addition by the generation units must be evaluated to adjust the difference between the measured and rated voltage. For the DN with V^{dc} nodes, a breakpoint matrix must be constructed. In the DS depicted in Figure 3.8, converter A operates at α while converter B operates at D. The simplified diagram for calculating $\mathbf{V}^{dc}$ breakpoint injection is presented in Figure 3.9. The detailed process for calculating $\mathbf{V}^{dc}$ breakpoint injection is explained below.
VDC = Set of all V^{dc} buses

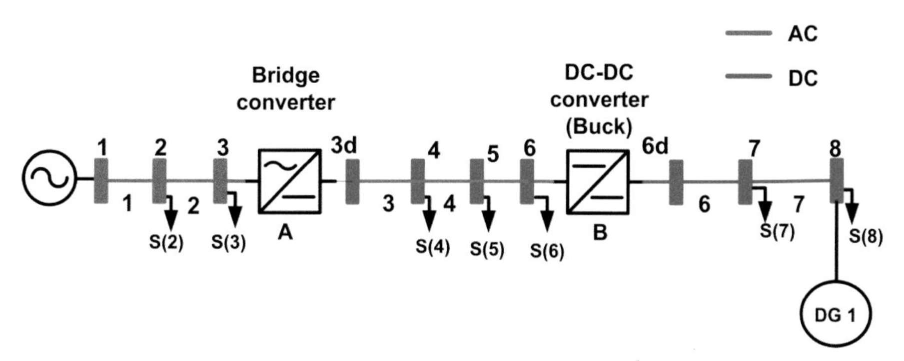

Figure 3.8 One-line diagram of DN with V^{dc} bus.

$VDC_W = V^{dc}\ W^{th}$ element
v^{dc} = total number of V^{dc} bus
$W = 1, 2........\ V^{dc}$

For the DS depicted in Figure 3.9, the V^{dc} breakpoint matrix can be written as follows:

$$[|Z(1)|\cos(\alpha)D^2 + |Z(2)|\cos(\alpha)D^2 + |\ R(3)\ |D^2 + |\ R(4)\ |D^2 + |\ R(5)\ |\ D^2 + (|R(6)| + |R(7)|)] \times \left[\Delta I^r\left(VDC_1\right)^t\right] = \left[\left|V\left(VDC_1''\right)^t\right| - \left|V\left(VDC_1\right)^t\right|\right] \tag{3.43}$$

In this case,

$$\Delta I^q(VDC_1) = 0$$

VDC = Set of all PV buses = $\{8\}$
$\text{v}^{dc} = 2 =$ Total number of V^{dc} buses
$W = 1$
$VDC_1 =$ 1st element of set VDC = 8
$\Delta V(VDC_1)^t$ = Voltage magnitude mismatch at the bus VDC_1 associated with V^{dc} DG number 1 at the start of iteration $t = V(VDC_1")^t - V(VDC_1)^t$,
$V(VDC_1")^t$ = Voltage magnitude (specified) at bus PVB_1 associated with V^{dc} DG number 1.

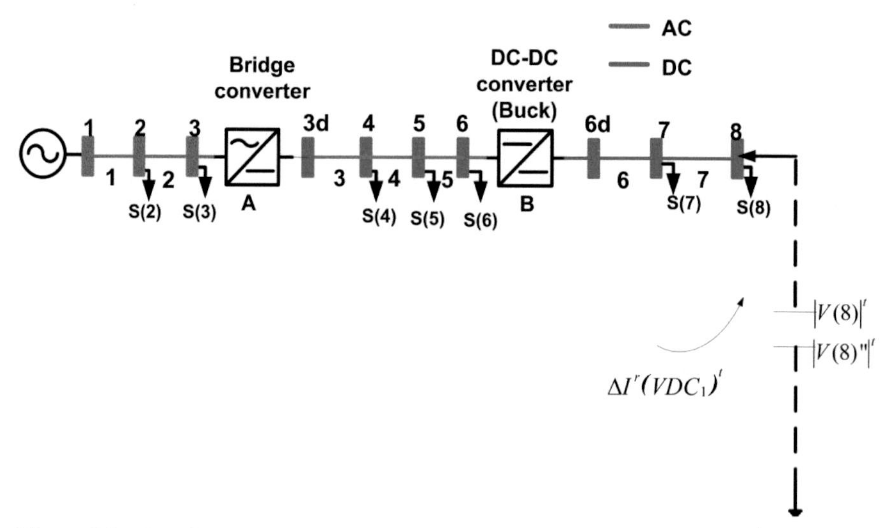

Figure 3.9 One-line representation of the DN depicted in Figure 3.8 for V^{dc} breakpoint injection calculations.

$V(VDC_1)^t$ = Voltage calculated (starting of iteration number t) at bus VDC_1 associated with V^{dc} DG number 1.
$\Delta I^r(VDC_1)^t$ = V^{dc} real current addition at the bus-VDC_1

The following equation holds under the presumption that all bus voltages are near 1.0 p.u. and ∅ is minimum.

$$\Delta I^r(VDC_{\mathrm{W}})^t = \Delta P(VDC_{\mathrm{W}})^t, \tag{3.44}$$

$\Delta P(VDC_W)^t$ = V^{dc} breakpoint real power addition at the bus-VDC_1 at the starting of tth iteration

Thus, eqn (3.43) can be modified as:

$$\begin{aligned}&\left[|Z(1)|\cos(\alpha)D^2 + |Z(2)|\cos(\alpha)D^2 + |R(3)|\,D^2 + |R(4)|\,D^2 + |R(5)|\,D^2\right.\\&\left.+(|R(6)| + |R(7)|)\right] \times \left[\Delta P(VDC_1)^t\right] = \left[\left|V(VDC_1")^t\right| - \left|V(VDC_1)^t\right|\right]\end{aligned} \tag{3.45}$$

The following equations describe the net active power added by the V^{dc}-type DG.

$$TP(VDC_1)^t = \sum_{x=1}^{t-1} \Delta P(VDC_1)^x + \Delta P(VDC_1)^t. \tag{3.46}$$

3.3.2 Case 2: The AC-DC distribution system with two V^{dc} nodes

An AC-DC DS illustrated in Figure 3.10, where DGs are designed as **V**^{dc} node/bus is supplying power at buses 8 and 5.

The exact quantity of real power addition by the generating units must be evaluated to adjust the difference between the measured and stated voltage. The breakpoint matrix has been constructed to sustain the required voltage at V^{dc} node. The simplified diagram for calculating **V**^{dc} breakpoint injections

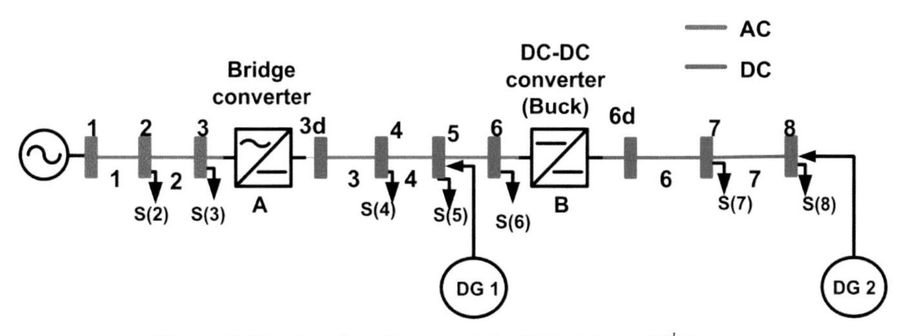

Figure 3.10 One-line diagram of the DN with two V^{dc} buses.

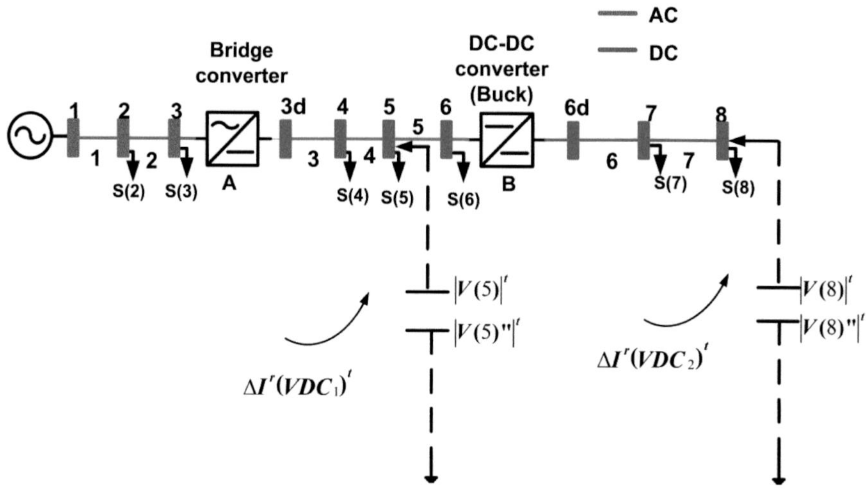

Figure 3.11 One-line representation of the DN depicted in Figure 3.10 for V^{dc} breakpoints injection calculation.

is illustrated in Figure 3.11. The detailed procedure for calculating $\mathbf{V}^{dc}$ breakpoint injections is exemplified below.

In the DS depicted in Figure 3.10, converter A operates at α while converter B operates at a D. For the DS depicted in Figure 3.10, the V^{dc} breakpoint matrix equation can be expressed as follows:

$$\begin{bmatrix} \begin{matrix}(|Z(1)|+|Z(2)|)\cos(\alpha)\\ +|Z(3)|+|Z(4)|\end{matrix} & \begin{matrix}(|Z(1)|+|Z(2)|)\cos(\alpha)D\\ +|Z(3)|D+|Z(4)|D\end{matrix} \\ \begin{matrix}(|Z(1)|+|Z(2)|)\cos(\alpha)D\\ +|Z(3)|D+|Z(4)|D\end{matrix} & \begin{matrix}(|Z(1)|+|Z(2)|)\cos(\alpha)D^2+(|Z(3)|\\ +|Z(4)|+|Z(5)|)D^2+|Z(6)|+|Z(7)|\end{matrix} \end{bmatrix}$$
$$\times \begin{bmatrix}\Delta I^r(VDC_1)^t\\ \Delta I^r(VDC_2)^t\end{bmatrix} = \begin{bmatrix}\left|V(VDC_1")^t\right| - \left|V(VDC_1)^t\right|\\ \left|V(VDC_2")^t\right| - \left|V(VDC_2)^t\right|\end{bmatrix} \tag{3.47}$$

In this particular case,

$$\Delta I^q(VDC_1) = \Delta I^q(VDC_2) = 0$$

VDC = set of all PV buses = $\{5,8\}$
$\text{v}^{dc} = 2$ = total number of V^{dc} buses
$W = 2$
VDC_1 = 1st element of set VDC = 5

$VDC_2 = $ 2nd element of set VDC = 8

$\Delta V(VDC_1)^t =$ Voltage magnitude mismatch at the bus VDC_1 associated with V^{dc} DG number 1 at the start of iteration $t = V(VDC_1")^t - V(VDC_1)^t$

$V(VDC_1")^t =$ Voltage magnitude (specified) at bus VDC_1 associated with V^{dc} DG number 1

$V(VDC_1)^t =$ Voltage calculated (beginning of iteration number t) at bus VDC_1 associated with V^{dc} DG number 1

$\Delta V(VDC_2)^t =$ Voltage magnitude mismatch at the bus VDC_2 associated with V^{dc} DG number 2 at the start of iteration $t = V(VDC_2")^t - V(VDC_2)^t$

$V(VDC_2")^t =$ Voltage magnitude (specified) at bus VDC_2 associated with V^{dc} DG number 2

$V(VDC_2)^t =$ Voltage calculated at bus VDC_1 associated with V^{dc} DG number 2

$\Delta I^r(VDC_1)^t =$ Additional V^{dc} breakpoint current addition at bus VDC_1

$\Delta I^r(VDC_2)^t =$ Additional V^{dc} breakpoint current addition at bus VDC_2

From the perspective of V^{dc} breakpoint real power injection, the previous equation for determining V^{dc} breakpoint current injection might be adjusted. The following equation applies under the hypothesis that all bus voltages are near 1 p.u. and $\emptyset$ is minimum.

$$\Delta I^r(VDC_W)^t = \Delta P(VDC_W)^t. \tag{3.48}$$

Thus, eqn (3.47) can be modified as:

$$\begin{bmatrix} (|Z(1)|+|Z(2)|)\cos(\alpha) & (|Z(1)|+|Z(2)|)\cos(\alpha)D \\ +|Z(3)|+|Z(4)| & +|Z(3)|D+|Z(4)|D \\ (|Z(1)|+|Z(2)|)\cos(\alpha)D & (|Z(1)|+|Z(2)|)\cos(\alpha)D^2+(|Z(3)| \\ +|Z(3)|D+|Z(4)|D & +|Z(4)|+|Z(5)|)D^2+|Z(6)|+|Z(7)| \end{bmatrix} \tag{3.49}$$

$$\times \begin{bmatrix} \Delta P(VDC_1)^t \\ \Delta P(VDC_2)^t \end{bmatrix} = \begin{bmatrix} \left|V(VDC_1")^t\right| - \left|V(VDC_1)^t\right| \\ \left|V(VDC_2")^t\right| - \left|V(VDC_2)^t\right| \end{bmatrix},$$

where

$\Delta P(VDC_1)^t =$ Additional V^{dc} breakpoint real current addition at bus VDC_1

$\Delta P(VDC_2)^t =$ Additional V^{dc} breakpoint real power addition at bus VDC_2.

The following equations produce the initial iteration t of total real power injections from the V^{dc} type dispersed generations in Figure 3.10:

$$TP(VDC_1)^t = \sum_{x=1}^{t-1} \Delta P(VDC_1)^x + \Delta P(VDC_1)^t \tag{3.50}$$

$$TP(VDC_2)^t = \sum_{x=1}^{t-1} \Delta P(VDC_2)^x + \Delta P(VDC_2)^t. \tag{3.51}$$

3.3.3 Case 3: The AC-DC distribution system with one PV node

A DG represented by the PV bus adding power into bus 8 of the DS is outlined in Figure 3.12.

The proper quantity of reactive power addition by the generators must be established to mitigate the difference between obtained and prescribed voltage. The breakpoint matrix was built to sustain the desired voltage at $\mathbf{V}^{dc}$ node. The simplified diagram for calculating *PV* breakpoint injection is presented in Figure 3.13. The detailed mechanism for calculating $\mathbf{V}^{dc}$ breakpoint injection is shown below.

For DS illustrated in Figure 3.13, the breakpoint equation may be expressed as follows:

$$\begin{aligned}&[\{(|Z(1)|+|Z(2)|)\cos(\alpha)MI+(|Z(3)|+|Z(4)|+|Z(5)|)MI\}\cos(\theta_v)\times\cos(\phi)\\&+(|X(6)|+|X(7)|)]\times\left[\Delta I^q(PVB_1)^t\right]=\left[\left|V(PVB_1")^t\right|-\left|V(PVB_1)^t\right|\right],\end{aligned} \tag{3.52}$$

$PVB =$ Set of all PV buses
$PVB_K = K^{th}$ element of set PVB
$pvb =$ Total V^{dc} buses
$K = 1, 2, \ldots\ldots.. pvb.$

In this particular case,

$$\Delta I^r(PVB_1) = 0$$
$$PVB = \{8\}$$
$$pvb = 1 = \text{Total PV buses}$$
$$K = 1$$
$$PVB_1 = \text{1st element of set PVB} = 8.$$

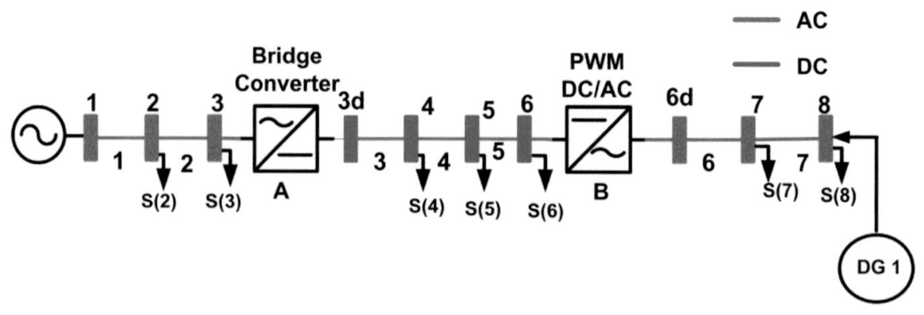

Figure 3.12 One-line diagram of DN with PV bus.

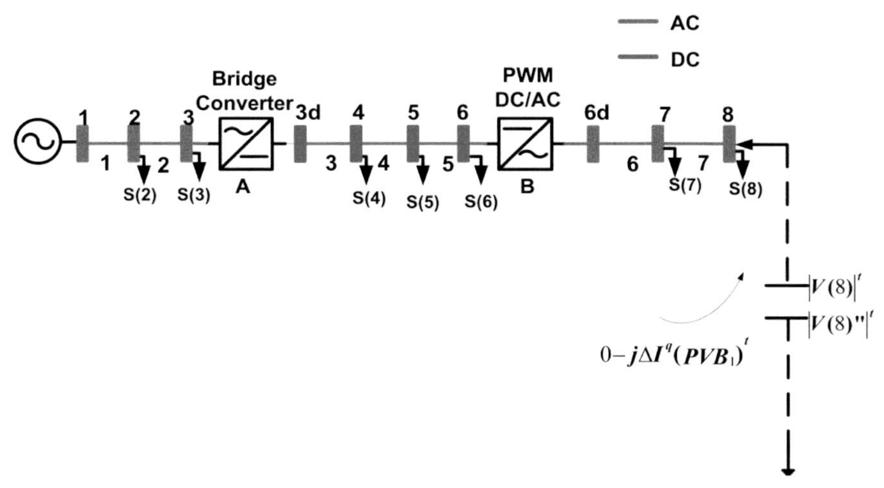

Figure 3.13 One-line representation of the DN depicted in Figure 3.12 for calculating *PV* breakpoint injection.

$\Delta V(PVB_1)^t$ = Voltage magnitude mismatch at the bus PVB_1 associated with PV DG number 1 at the starting of iteration t = $V(PVB_1")^t - V(PVB_1)^t$

$V(PVB_1")^t$ = Voltage magnitude (specified) at bus PVB_1 associated with PV DG number 1

$V(PVB_1)^t$ = Voltage calculated (starting of iteration number t) at bus PVB_1 associated with PV DG number 1

$\Delta I^q(PVB_1)^t$ = Additional PV breakpoint reactive current addition at PVB_1 bus in the starting of tth iteration

θ_v = Phase angle of the converter B terminal voltage.

The following equation for estimating *PV* breakpoint current addition may be altered to account for PV breakpoint reactive power injection. The following equation holds assuming that all bus voltages are near 1 p.u. and ∅ is minimum.

$$\Delta I^q(PVB_K)^t = \Delta Q(PVB_K)^t. \tag{3.53}$$

Thus, eqn (3.52) can be modified as:

$$\begin{aligned}&[\{(|Z(1)|+|Z(2)|)\cos(\alpha)MI+(|Z(3)|+|Z(4)|+|Z(5)|)MI\}\cos(\theta_v)\times\cos(\phi)\\&+(|X(6)|+|X(7)|)]\times\left[\Delta Q(PVB_1)^t\right]=\left[\left|V(PVB_1")^t\right|-\left|V(PVB_1)^t\right|\right]\end{aligned}, \tag{3.54}$$

where
$\Delta Q(PVB_1)^t$ = Additional *PV* breakpoint reactive power addition at PVB_1 bus in the starting of *t*th iteration.

The DG in Figure 3.12 injects reactive power at the start of *t*th iteration as demonstrated by the equation:

$$TQ(PVB_1)^t = \sum_{x=1}^{t-1} \Delta Q(PVB_1)^x + \Delta Q(PVB_1)^t \tag{3.55}$$

3.3.4 Case 4: The AC-DC distribution system with both PV and V^{dc} buses

Consider the DS outlined in Figure 3.14, DG represented as V^{dc}-bus and PV-bus inject power at bus 8 and 2, respectively. For the DG modelled as V^{dc} bus, the generation units must inject/withdraw the right amount of active power/current to compensate for the voltage difference (V^{dc} bus voltage). To compensate for voltage variance in DG modelled as a PV bus, generation units must inject/withdraw the right amount of reactive power/current (PV bus voltage). To calculate the injections by both DGs a common breakpoint matrix has been developed. The detailed procedure is shown below.

For the DS illustrated in Figure 3.14, the common breakpoint matrix for calculating V^{dc} and PV breakpoint injections is given by the following equation:

$$\begin{bmatrix} |Z(1)| & |Z(1)| \\ |Z(1)|\cos(\alpha) & (|Z(1)|+|Z(2)|)\cos(\alpha)+|Z(3)|+|Z(4)| \end{bmatrix} \times \begin{bmatrix} \Delta I^q(PVB_1)^t \\ \Delta I^r(VDC_1)^t \end{bmatrix} = \begin{bmatrix} \left||V(PVB_1")^t| - |V(PVB_1)^t|\right| \\ |V(VDC_1")^t| - |V(VDC_1)^t| \end{bmatrix}. \tag{3.56}$$

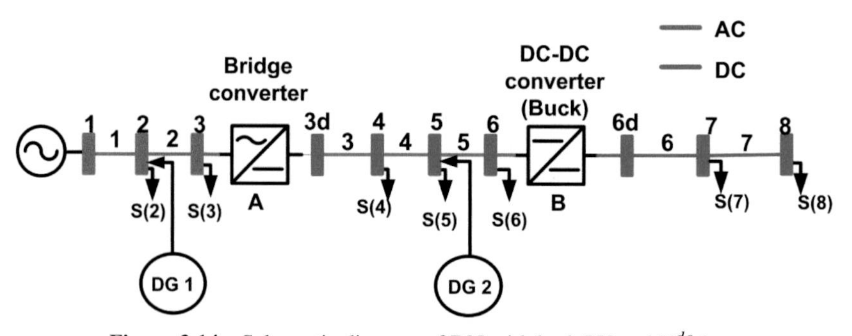

Figure 3.14 Schematic diagram of DN with both PV and V^{dc} bus.

In this particular case,

$$\Delta I^q(VDC_1) = \Delta I^q(VDC_2) = 0$$

$$\begin{aligned}VDC &= \text{Set of all } V^{dc}\text{-buses} = \{5\}\\ PVB &= \text{Set of all PV-buses} = \{2\}\\ VDC &= \text{Set of all V}^{dc}\text{-buses} = \{5\}\\ \text{vdc} = 1 &= \text{Total V}^{dc} \text{ buses}\\ pvb = 1 &= \text{Total PV buses}\\ K &= 1 \text{ for PVB}\\ W &= 1 \text{ for VDC}\\ VDC_1 &= \text{1st element of set VDC} = 5\\ PVB_1 &= \text{1st element of set PVB} = 2.\end{aligned}$$

$\Delta V(VDC_1)^t$ = Voltage magnitude mismatch at the bus VDC_1 associated with V^{dc} DG number 1 at the start of iteration $t = V(VDC_1")^t - V(VDC_1)^t$
$V(VDC_1")^t$ = Voltage magnitude (specified) at bus VDC_1 associated with V^{dc} DG number 1
$V(VDC_1)^t$ = Voltage calculated (starting of iteration number t) at bus VDC_1 associated with V^{dc} DG number 1
$\Delta I^r(VDC_1)^t$ = Additional V^{dc} breakpoint real current addition at VDC_1 bus at the starting of tth iteration.
$\Delta V(PVB_1)^t$ = Voltage magnitude mismatch at the bus PVB_1 associated with *PV* DG number 1 at the starting of iteration $t = V(PVB_1")^t - V(PVB_1)^t$
$V(PVB_1")^t$ = Voltage magnitude (specified) at bus PVB_1 associated with *PV* DG number 1
$V(PVB_1)^t$ = Voltage calculated (starting of iteration number t) at bus PVB_1 associated with *PV* DG number 1
$\Delta I^q(PVB_1)^t$ = Additional *PV* breakpoint reactive current addition at PVB_1 bus at the starting of tth iteration

Under the conditions that all bus voltages are near 1 p.u. and $\emptyset$ is minimum, the supporting equation holds:

$$\Delta I^r(VDC_W)^t = \Delta P(VDC_W)^t, \tag{3.57}$$

$$\Delta I^q(PVB_K)^t = \Delta Q(PVB_K)^t. \tag{3.58}$$

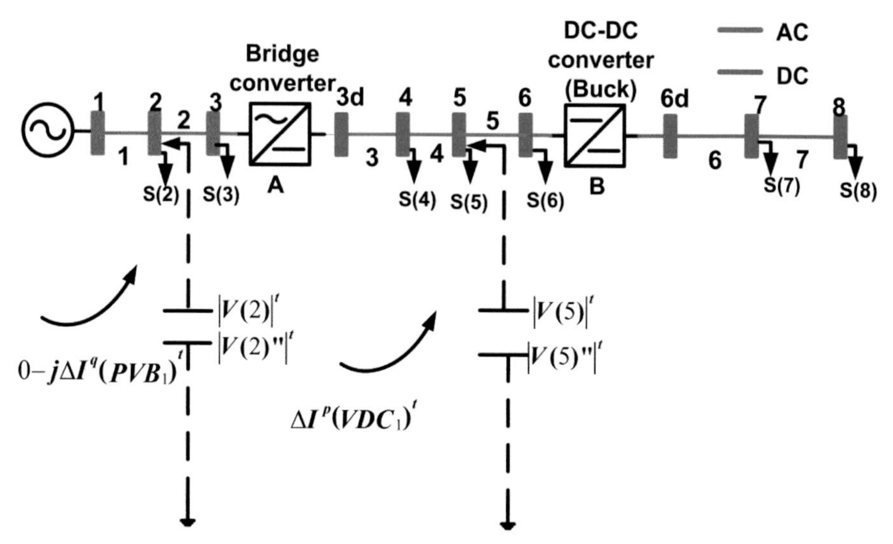

Figure 3.15 One-line representation of DN depicted in Figure 3.14 for calculating *PV* and V^{dc} breakpoint injections.

Thus, eqn (3.56) can be modified as:

$$\begin{bmatrix} |Z(1)| & |Z(1)| \\ |Z(1)|\cos(\alpha) & (|Z(1)|+|Z(2)|)\cos(\alpha)+|Z(3)|+|Z(4)| \end{bmatrix} \times \begin{bmatrix} \Delta Q(PVB_1)^t \\ \Delta P(VDC_1)^t \end{bmatrix} = \begin{bmatrix} \left||V(PVB_1")^t| - |V(PVB_1)^t|\right| \\ |V(VDC_1")^t| - |V(VDC_1)^t| \end{bmatrix} \tag{3.59}$$

The following equations give the total real and reactive power addition by V^{dc} and PV DG (Figure 3.15):

$$TP(VDC_1)^t = \sum_{x=1}^{t-1} \Delta P(VDC_1)^x + \Delta P(VDC_1)^t, \tag{3.60}$$

$$TQ(PVB_1)^t = \sum_{x=1}^{t-1} \Delta Q(PVB_1)^x + \Delta Q(PVB_1)^t. \tag{3.61}$$

The complete steps for load flow computations are provided in Figure 3.16.

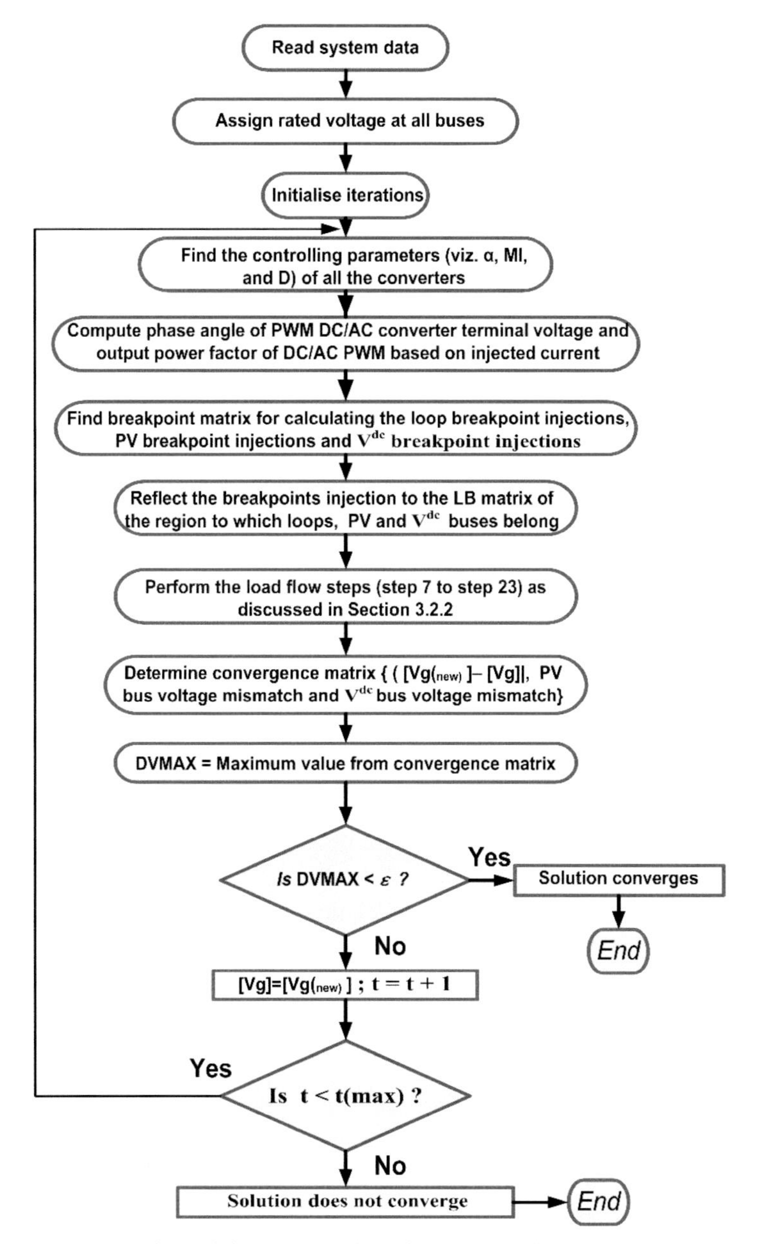

Figure 3.16 LF algorithm for DS (including DG).

3.4 Results

This section demonstrates the findings that were utilized to test the accuracy of the proposed LF model. Using MATLAB software on a desktop with Intel core-i7, 3.4 GHz, 64-bit, and 8 GB RAM, the algorithm was tested. In this portion, the outcomes of the method for calculating load flow and a comparison of its effectiveness with that of algorithms already in use are given.

3.4.1 Test system 1: 10 bus AC-DC distribution network

The initial study considered for the DN is a 10-bus network, as shown in Figure 3.17. Table 3.2 and Table 3.3 show the information about the loads and generators for each bus. The information about the lines is given in Table 3.2. One line-commutated converter is linked to buses 3^{ac} and 4 in this network. This converter has a minimum firing angle of 0°, and its commutation reactance is ignored. A three-phase bridge converter maintains voltage at 0.9750 p.u. at bus 4. The BBC is linked to buses 6^{dc} and 7 to regulate the voltage at bus 7–0.9801 p.u. In Table 3.4, information about different types of converters is presented. Base-voltage = 12.66 kV (AC-side), base-voltage = 17.09 kV (DC-side) and base-MVA = 100. All converters are assumed loss less (100% efficient). (Tolerance–10^{-6})

Table 3.5 provides a detailed breakdown of the LF results for the mentioned DS using the suggested methodology, the BIBC (bus injection to branch current) technique, and the LF results given by PSCAD. According to the data that was collected, there is a maximum discrepancy of 0.0003 between the bus voltages that were estimated using the load flow methodology that was suggested and the LF model that was simulated in PSCAD. The suggested LF approach and the BIBC LF model have a maximum difference of 0.0002 volts in the bus voltages that they compute using their respective methods. The proposed methodology takes 20 ms more than the

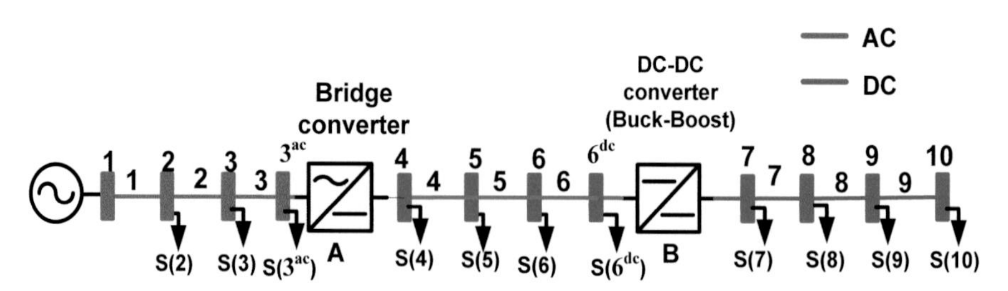

Figure 3.17 Schematic representation of 10-bus AC-DC DN.

Table 3.2 Data of hypothetical 10-bus system shown in Figure 3.17.

BN	SN	RN	Impedance	Load data (S(RN)) (kVA/kW)
1	1	2	0.84111 + 0.82271i	44.1 + 44.98i
2	2	3	0.84111 + 0.82271i	44.1 + 44.98i
3	3	3ac	0.84111 + 0.82271i	0
-	3ac	4	LCC-converter	264.6
4	4	5	1.6822	132.3
5	5	6	1.6822	44.1
6	6	6dc	1.6822	0
-	6dc	7	DC-DC boost converter	44.1
7	7	8	1.6822	44.1
9	8	9	1.6822	44.1
10	9	10	1.6822	44.1

Table 3.3 Characteristics of controlled generators.

Generator-number (types)	Bus location	Real power injected (kW)		Reactive power injected (kVAr)		Voltage (p.u.)
		Max.	Min.	Max.	Min.	
G1 (V^{dc})	5	500	50	-	-	0.9745
G2 (V^{dc})	10	500	50	-	-	0.9782
G4 (PQ)	3	20	20	10	10	-

Table 3.4 Characteristics of converters.

Types of converters	Bus location	Control algorithm	Control limit
$3-\Phi$ full bridge converter	3^{ac} and 4	Fixed DC voltage	$\alpha(\min) = 7^{\circ}$ $\alpha(\max) = 18^{\circ}$
DC-DC converter	6^{dc} and 7	Fixed voltage	$D(\min) = 0.1$ $D(\max) = 0.7$

Table 3.5 Voltage profile for hypothetical 10-bus AC-DC DS (without DGs).

Bus-number	Magnitude of voltage (p.u.)			Phase (Radian)		
	PM	BIBC	PSCAD	PM	BIBC	PSCAD
1	1.0000	1.0000	1.0000	0.0000	0.0000	0.0000
2	0.9965	0.9966	0.9964	−0.0037	−0.0038	−0.0038
3	0.9936	0.9935	0.9936	−0.0075	−0.0090	−0.0091
3^{ac}	0.9910	0.9909	0.9908	−0.0110	−0.0112	−0.0111
4	0.9750	0.9750	0.9750	-	-	
5	0.9729	0.9729	0.9729	-	-	
6	0.9716	0.9716	0.9715	-	-	
6^{dc}	0.9705	0.9705	0.9705	-	-	
7	0.9801	0.9801	0.9801	-	-	
8	0.9780	0.9779	0.9778	-	-	
9	0.9767	0.9765	0.9765	-	-	
10	0.9756	0.9755	0.9754	-	-	

BIBC method, which takes 200 ms. The reliability and precision of the newly devised approach are shown by a comparison of the outcomes produced using the BIBC technique, PSCAD, and the suggested methodology.

The BIBC matrix method has applicability limited only to simple AC-DC distribution networks. This approach (BIBC) has not considered the effect of numerous models of DGs. The suggested LF algorithm has resolved the limitations of the existing methodology and can handle mathematical models of DGs. An experiment was carried out on DN illustrated in Figure 3.17 in the presence of DGs to show that the suggested strategy is efficient.

Table 3.6 Voltage profile for hypothetical 10-bus AC-DC DS (DGs included).

Bus-number	Magnitude of voltage (p.u.)		Phase (radian)	
	PM	PSCAD	PM	PSCAD
1	1.0000	1.0000	0.0000	0.0000
2	0.9977	0.9976	−0.0021	−0.0022
3	0.9958	0.9957	−0.0044	−0.0044
3^{ac}	0.9943	0.9942	−0.0066	−0.0067
4	0.9750	0.9750	-	-
5	0.9745	0.9744	-	-
6	0.9736	0.9735	-	-
6^{dc}	0.9730	0.9729	-	-
7	0.9801	0.9801	-	-
8	0.9795	0.9795	-	-
9	0.9787	0.9788	-	-
10	0.9782	0.9781	-	-

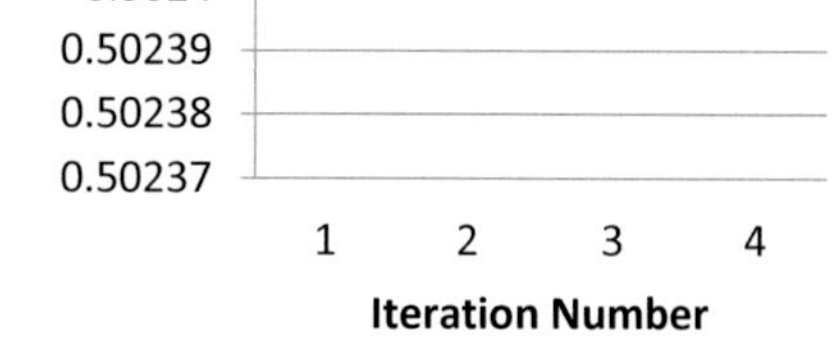

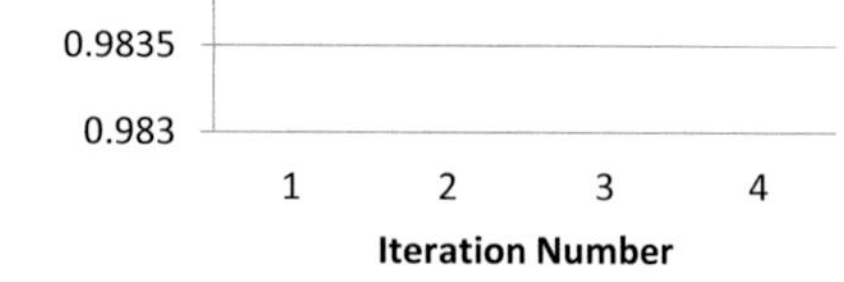

Figure 3.18 Variation of control variables of the converters in the 10 bus AC-DC DN (Figure 3.17) without DG during load-flow iteration: (a) changes in the duty-ratio for converter B, (b) changes in the cos(α) of the converter A.

Table 3.7 Distributed generator's load-flow results.

Generator-number (types)	Real power injected (kW)		Reactive power injected (kVAr)		Voltage (p.u.)
	PM	PSCAD	PM	PSCAD	
G1 (V^{dc})	80.47	80.48	-	-	0.9745
G2 (V^{dc})	173.90	174	-	-	0.9782
G4 (PQ)	20	20	10	10	-

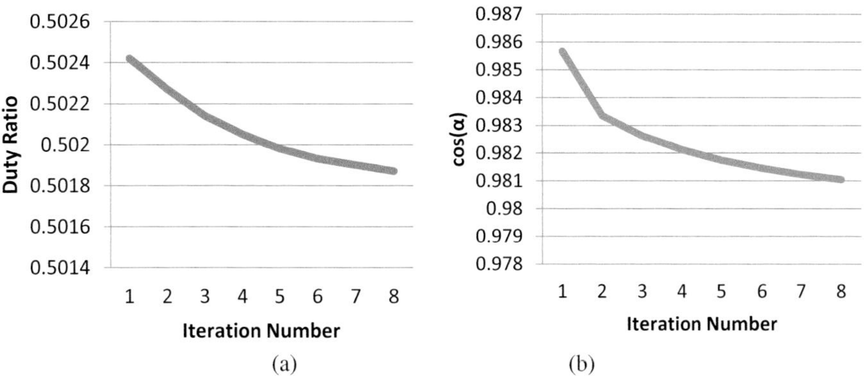

Figure 3.19 Variation of control variables of the converters in the network (In Figure 3.17) with DG during load flow iteration: (a) changes in the duty ratio for converter B, (b) changes in the cos(α) of the converter A.

The LF results for the DS with DGs using the suggested algorithm and the LF outcomes furnished by PSCAD is explicitly displayed in Table 3.6 and Table 3.7. The efficiency and precision of the established technique are shown by comparing the outcomes acquired from the suggested algorithm with those generated by the PSCAD application.

The DC-DC BBC D and α changes during the LF iterations of the suggested technique are illustrated in Figure 3.18 (for base network) and Figure 3.19 (base network with the inclusion of DGs).

3.4.2 Test system 2: 15 bus AC-DC distribution network

The 15-bus DN is the second test system (Figure 3.20). Table 3.8 and Table 3.9 provide details about the loads and generators respectively. The line data information is outlined in Table 3.8.

One line-commutated converter combines buses 3^{ac} and 4 in this network. This converter has a 0° firing angle and no commutation reactance.

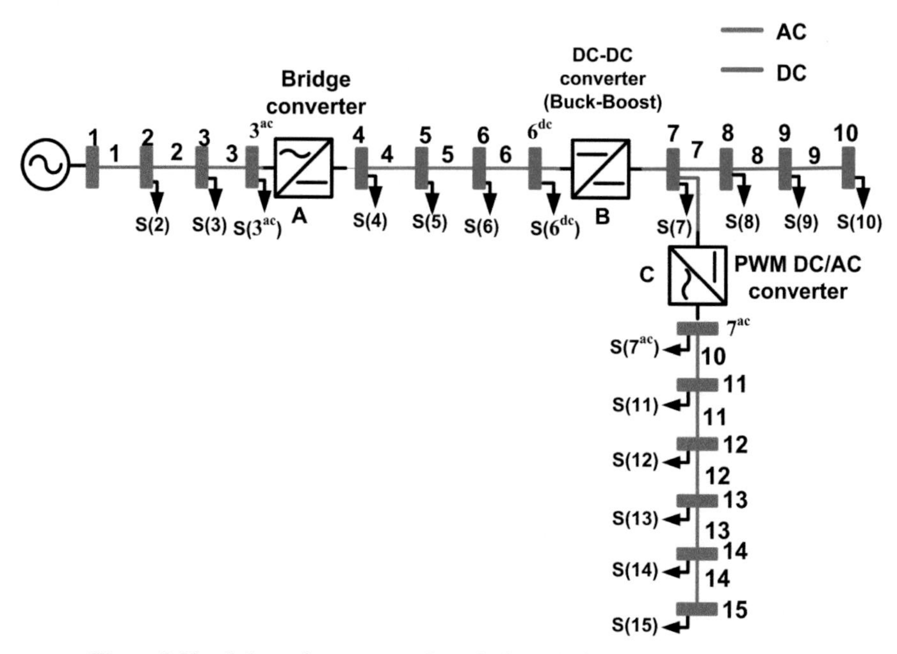

Figure 3.20 Schematic representation of a hypothetical 15-bus AC-DC DN.

Three-phase bridge converters manage bus 4 voltage at 0.9750 p.u. The DC-DC BBC around buses 6^{dc} and 7 establish a constant voltage of 0.9801 p.u. at bus 7. The converter between buses 7 and 7^{ac} maintains a 0.9701 p.u. bus voltage. Table 3.10 lists converter data.

The LF outcomes for the aforesaid DS using the suggested technique, the BIBC technique, and the LF outcomes produced by PSCAD are reported precisely in Table 3.11. The maximum difference between the bus voltages estimated by the suggested LF method and the LF model tested in PSCAD is not more than 0.0003. The proposed LF algorithm and the BIBC LF model differ by 0.0002 when calculating bus voltages. The suggested and the BIBC technique take 30 and 250 ms to execute respectively. When the results from the BIBC method, the PSCAD software, and the suggested algorithm are compared, it highlights the efficiency and accuracy of the developed method.

The LF outcomes for the previously specified DS (with DGs) utilizing the suggested method and the solution generated by the PSCAD application are reported in an elaborative manner in Table 3.12 and Table 3.13. According to the data that was collected, there is a maximum difference of 0.0001 between the bus voltages determined by the suggested LF approach and the LF model

Table 3.8 Data of hypothetical 15-bus system shown in Figure 3.20.

BN	SN	RN	Impedance	Load data (S(RN)) (kVA/kW)
1	1	2	0.84111 + 0.82271i	44.1 + 44.98i
2	2	3	0.84111 + 0.82271i	44.1 + 44.98i
3	3	3^{ac}	0.84111 + 0.82271i	0
-	3^{ac}	4	LCC converter	264.6
4	4	5	1.6822	132.3
5	5	6	1.6822	44.1
6	6	6^{dc}	1.6822	0
-	6^{dc}	7	DC-DC boost converter	44.1
7	7	8	1.6822	44.1
8	8	9	1.6822	44.1
9	9	10	1.6822	44.1
-	7	7^{ac}	DC-AC PWM converter	0
10	7^{ac}	11	0.84111 + 0.82271i	44.1 + 44.98i
11	11	12	0.84111 + 0.82271i	44.1 + 44.98i
12	12	13	0.84111 + 0.82271i	132.3 +134.96i
13	13	14	0.84111 + 0.82271i	88.20 + 89.97i
14	14	15	0.84111 + 0.82271i	88.20 + 89.97i

Table 3.9 Characteristics of controlled generators.

Generator-number (types)	Location of buses	Real power injected (kW)		Reactive power injected (kVAr)		Voltage (p.u.)
		Max.	Min.	Max.	Min.	
G-1 (P)	5	60	60	-	-	-
G-2 (V^{dc})	10	500	50	-	-	0.9797
G-4(PV)	15	100	100	20	125	0.9600

Table 3.10 Characteristics of converters.

Types of converters	Bus location	Control algorithm	Control limit
3- Φ full bridge converter	3^{ac} and 4	Fixed DC voltage	$\alpha(\min) = 7^\circ$ $\alpha(\max) = 18^\circ$
DC-DC converter	6^{dc} and 7	Fixed Voltage	$D(\min) = 0.1$ $D(\max) = 0.7$
PWM DC-AC converter	7 and 7^{ac}	Fixed voltage	$MI(\min) = 0.5$ $MI(\max) = 1$

that was designed in PSCAD. The efficacy and reliability of the approach that was developed are shown by portraying the outcomes acquired from the suggested technique with those provided by PSCAD.

Table 3.11 Voltage profile for hypothetical 15-bus AC-DC DS (DGs not included).

Bus-number	Magnitude of voltage (p.u.)			Phase (Radian)		
	PM	BIBC	PSCAD	PM	BIBC	PSCAD
1	1.0000	1.0000	1.0000	0.0000	0.0000	0.0000
2	0.9929	0.9928	0.9929	−0.0047	−0.0046	−0.0049
3	0.9864	0.9863	0.9865	−0.0095	−0.0097	−0.0098
3^{ac}	0.9803	0.9801	0.9804	−0.0144	−0.0146	−0.0147
4	0.9750	0.9750	0.9751	-	-	
5	0.9705	0.9704	0.9707	-	-	
6	0.9668	0.9667	0.9669	-	-	
6^{dc}	0.9633	0.9632	0.9633	-	-	
7	0.9801	0.9801	0.9801	-	-	
8	0.9793	0.9792	0.9793	-	-	
9	0.9787	0.9787	0.9788	-	-	
10	0.9785	0.9784	0.9785	-	-	
7^{ac}	0.9701	0.9701	0.9701	−0.0102	−0.0102	−0.0102
11	0.9658	0.9659	0.9658	−0.0101	−0.0101	−0.0101
12	0.9619	0.9620	0.9619	−0.0100	−0.0100	−0.0100
13	0.9585	0.9586	0.9587	−0.0099	−0.0099	−0.0099
14	0.9566	0.9567	0.9567	−0.0099	−0.0099	−0.0099
15	0.9556	0.9557	0.9558	−0.0098	−0.0098	−0.0098

Table 3.12 Voltage profile for hypothetical 15-bus DS.

Bus-number	Voltage magnitude (p.u.)		Phase (Radian)	
	PM	PSCAD	PM	PSCAD
1	1.0000	1.0000	0.0000	0.0000
2	0.9942	0.9943	−0.0035	−0.0034
3	0.9889	0.9889	−0.0071	−0.0071
3^{ac}	0.9841	0.9840	−0.0107	−0.0107
4	0.9750	0.9750	-	-
5	0.9719	0.9719	-	-
6	0.9691	0.9690	-	-
6^{dc}	0.9667	0.9666	-	-
7	0.9801	0.9801	-	-
8	0.9797	0.9797	-	-
9	0.9796	0.9796	-	-
10	0.9797	0.9797	-	-
7^{ac}	0.9701	0.9701	−0.0075	−0.0075
11	0.9666	0.9665	−0.0072	−0.0072
12	0.9637	0.9636	−0.0069	−0.0069
13	0.9612	0.9611	−0.0066	−0.0066
14	0.9601	0.9601	−0.0063	−0.0063
15	0.9600	0.9600	−0.0061	−0.0061

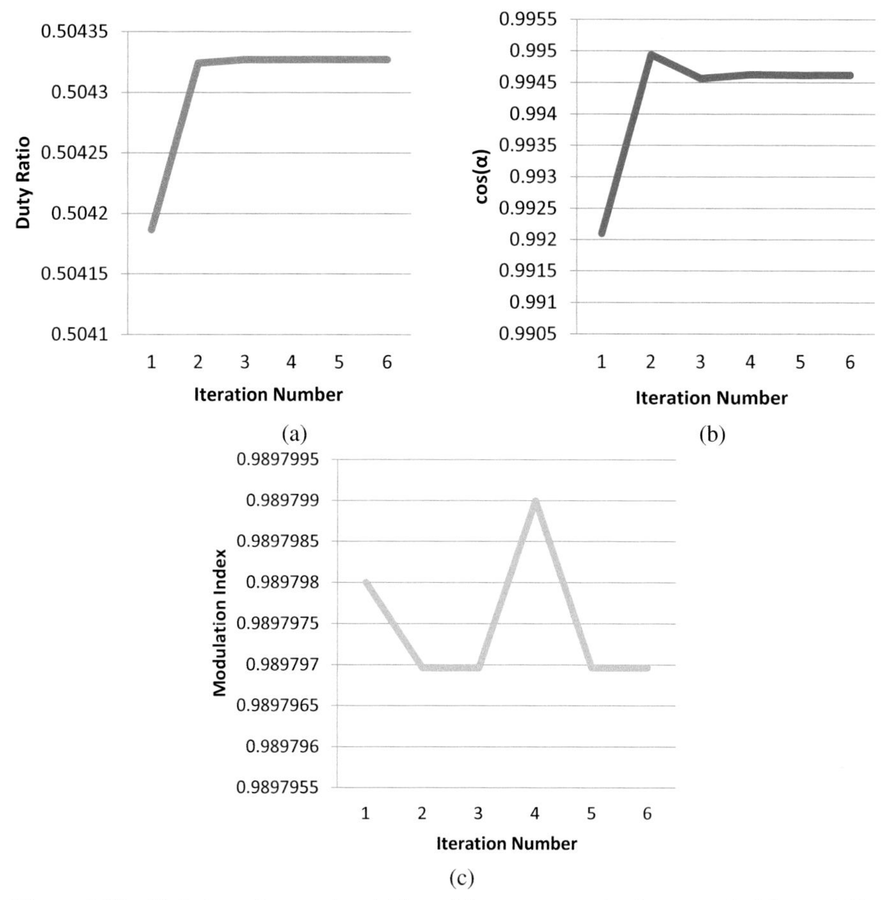

Figure 3.21 Variation of control variables of the converters in the network (Figure 3.20) without DG during load flow iteration: (a) duty ratio variation for converter B, (b) changes in the cosine of commutation angle of the converter A, (c) duty ratio variation of the converter C.

Table 3.13 Distributed generator load-flow results.

Generator-number (types)	**Real power injected (kW)**		**Reactive power injected (kVAr)**		**Voltage (p.u.)**
	PM	PSCAD	PM	PSCAD	
G-1(P)	60	60	-	-	-
G-2 (V^{dc})	65.29	65.28	-	-	0.9797
G-4(PV)	100	100	59.84	59.84	0.9600

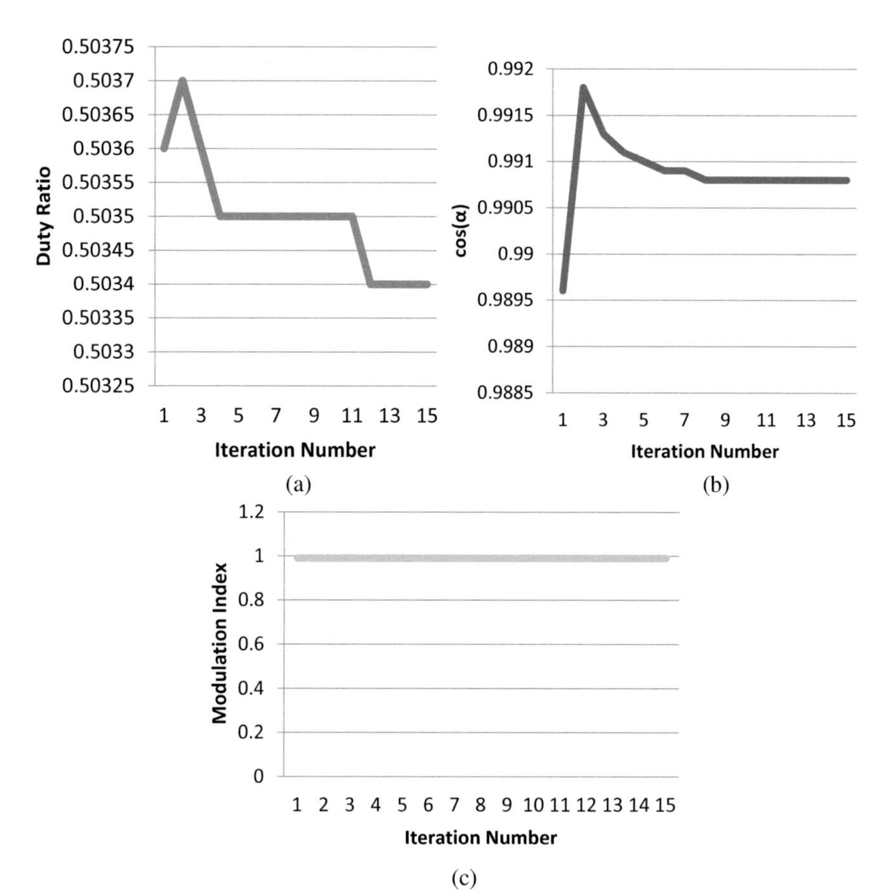

Figure 3.22 Variation of control variables of the converters in the network (Figure 3.20) with DG during load flow iteration: (a) duty ratio variation for converter B, (b) changes in the $\cos(\alpha)$ of the converter A, (c) duty ratio variation for converter C.

Variations in the DC-DC BBC duty cycle, the rectifier's firing angle $\cos(\alpha)$, and the PWM DC-AC converter's modulation index during the LF iterations of the suggested technique are displayed in Figure 3.21 and Figure 3.22 for the AC-DC DS depicted in Figure 3.20 (without and with DGs respectively).

The suggested scheme has been examined for a broad variation of $\frac{R}{X}$ ratio of lines corresponding to the 15-bus AC-DC radial network. The impact of $\frac{R}{X}$ deviation on the convergence characteristics is evaluated. The test outcomes illustrated in Figure 3.23 illustrate the convergence of the proposed technique.

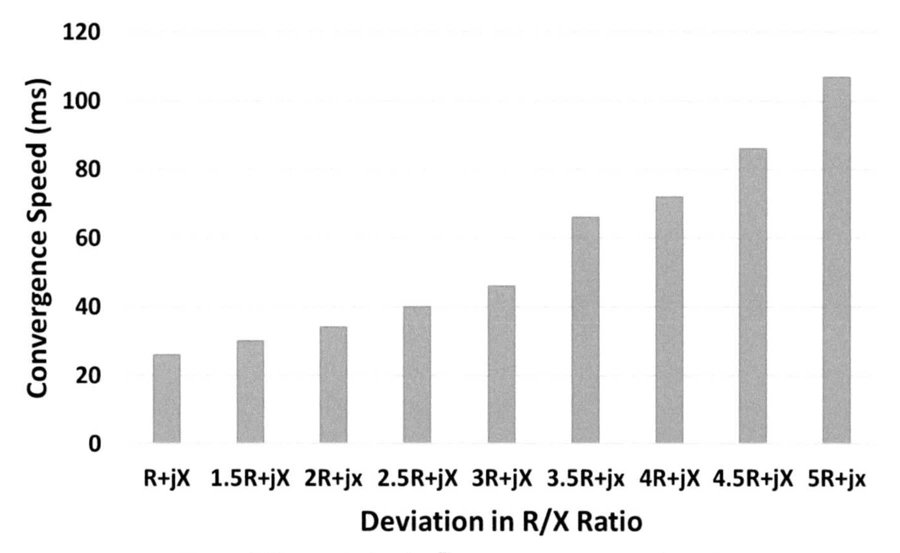

Figure 3.23 Deviation in $\frac{R}{X}$ vs. convergence speed graph.

3.5 Conclusion

A LF methodology for an AC-DC radial DS is provided in this chapter. A p.u. model of numerous power system converters is introduced to bridge the difference between an AC and DC DS. The feature of the converter model introduced allows us to carry out the LF of the DS. The different ways that power converters can operate have also been considered to show the effectiveness of the suggested approach. The sensitivity and breakpoint matrix have been designed to access the LF solution of DN with PV and V^{dc} bus. The suggested method is computationally effective due to the concise algebraic operation and effective search technique. In the absence of a standard DS, fictitious test cases (10-bus and 15-bus AC-DC DS) have been considered and the effectiveness of the LF algorithm is examined. The suggested model's efficacy has been explored in comparison to the LF solution generated by the PSCAD and BIBC methods. According to the test results, the proposed LF technique can offer concise results while also providing agility and quickness. The proposed load flow algorithm's ability to converge has been tested, and the system $\frac{R}{X}$ has been transformed to test the method's ruggedness. For a broad range of $\frac{R}{X}$, this load flow method converges. The convergence speed will, however, slow down as the $\frac{R}{X}$ rises.

References

[1] G. F. Reed, B. M. Grainger, A. R. Sparacino and Z. -H. Mao, “Ship to Grid: Medium-Voltage DC Concepts in Theory and Practice,” IEEE Power and Energy Magazine, vol. 10, no. 6, pp. 70-79, Nov.-Dec. 2012.

[2] T. Adefarati and R. C. Bansal, “Integration of renewable distributed generators into the distribution system: A review,” IET Renew. Power Gener., vol. 10, no. 7, pp. 873-884, Jul. 2016.

[3] C. Zhao, S. Dong, C. Gu, F. Li, Y. Song and N. P. Padhy, “New Problem Formulation for Optimal Demand Side Response in Hybrid AC-DC Systems,” IEEE Transactions on Smart Grid, vol. 9, no. 4, pp. 3154-3165, July 2018

[4] N. Kinhekar, N. P. Padhy, F. Li and H. O. Gupta, “Utility Oriented Demand Side Management Using Smart AC and Micro DC Grid Cooperative,” IEEE Transactions on Power Systems, vol. 31, no. 2, pp. 1151-1160, March 2016.

[5] S. Messalti, S. Belkhiat, S. Saadate and D. Flieller, “A new approach for load flow analysis of integrated AC-DC power systems using sequential modified Gauss–Seidel methods,” Eur. Trans. Electr. Power, vol. 22, pp. 421-432, 2012.

[6] J. Beerten, S. Cole and R. Belmans, “Generalized Steady-State VSC MTDC Model for Sequential AC-DC Power Flow Algorithms,” IEEE Transactions on Power Systems, vol. 27, no. 2, pp. 821-829, May 2012

[7] C. Liu, B. Zhang, Y. Hou, F. F. Wu and Y. Liu, “An Improved Approach for AC-DC Power Flow Calculation with Multi-Infeed DC Systems,” IEEE Transactions on Power Systems, vol. 26, no. 2, pp. 862-869, May 2011

[8] M. Baradar and M. Ghandhari, “A Multi-Option Unified Power Flow Approach for Hybrid AC-DC Grids Incorporating Multi-Terminal VSC-HVDC,” IEEE Transactions on Power Systems, vol. 28, no. 3, pp. 2376-2383, Aug. 2013

[9] M. E. El-Hawary and S. T. Ibrahim, “A new approach to AC-DC load flow analysis,” Elect. Power Syst. Res., vol. 33, no. 3, pp. 193-200, 1995.

[10] S. Iwamoto and Y. Tamura, “A Load Flow Calculation Method for Ill-Conditioned Power Systems,” IEEE Transactions on Power Apparatus and Systems, vol. PAS-100, no. 4, pp. 1736-1743, April 1981.

[11] Jen-Hao Teng and Chuo-Yean Chang, "A novel and fast three-phase load flow for unbalanced radial distribution systems," IEEE Transactions on Power Systems, vol. 17, no. 4, pp. 1238-1244, Nov. 2002.
[12] B. Venkatesh, A. Dukpa and L. Chang, "An accurate voltage solution method for radial distribution systems," Canadian Journal of Electrical and Computer Engineering, vol. 34, no. 1/2, pp. 69-74, Winter-Spring 2009.
[13] R. D. Zimmerman and Hsiao-Dong Chiang, "Fast decoupled power flow for unbalanced radial distribution systems," IEEE Transactions on Power Systems, vol. 10, no. 4, pp. 2045-2052, Nov. 1995.
[14] A. Losi and M. Russo, "Object-oriented load flow for radial and weakly meshed distribution networks," IEEE Transactions on Power Systems, vol. 18, no. 4, pp. 1265-1274, Nov. 2003.
[15] T. Ochi, D. Yamashita, K. Koyanagi and R. Yokoyama, "The development and the application of fast decoupled load flow method for distribution systems with high R/X ratios lines," 2013 IEEE PES Innovative Smart Grid Technologies Conference (ISGT), Washington, DC, USA, 2013, pp. 1-6.
[16] P. Aravindhababu and R. Ashokkumar, "A Fast Decoupled Power Flow for Distribution Systems," Electric Power Components and Systems, vol. 31, no. 9, pp. 932-940, 2008.
[17] O. L. Tortelli, E. M. Lourenço, A. V. Garcia and B. C. Pal, "Fast Decoupled Power Flow to Emerging Distribution Systems via Complex pu Normalization," IEEE Transactions on Power Systems, vol. 30, no. 3, pp. 1351-1358, May 2015.
[18] S. Ghosh and D. Das, "Method for load-flow solution of radial distribution networks," Proc. IEE—Gener. Transm. Distrib., vol. 146, no. 6, pp. 641-648, 1999.
[19] G. W. Chang, S. Y. Chu and H. L. Wang, "An Improved Backward/Forward Sweep Load Flow Algorithm for Radial Distribution Systems," IEEE Transactions on Power Systems, vol. 22, no. 2, pp. 882-884, May 2007.
[20] T. Alinjak, I. Pavić and M. Stojkov, "Improvement of backward/forward sweep power flow method by using modified breadth-first search strategy," IET Gener. Transm. Distrib., vol. 11, no. 1, pp. 102-109, Jan. 2017.
[21] M. R. Shakarami, H. Beiranvand, A. Beiranvand and E. Sharifipour, "A recursive power flow method for radial distribution networks: Analysis

solvability and convergence," Int. J. Elect. Power Energy Syst., vol. 86, pp. 71-80, Mar. 2017.
[22] Jen-Hao Teng, "A direct approach for distribution system load flow solutions," IEEE Transactions on Power Delivery, vol. 18, no. 3, pp. 882-887, July 2003.
[23] G. Abad, J. Lopez, M.A. Roderiguez, M. Marroyo, G. Iwanski, "Doubly fed induction machine: modeling and control for wind energy generation," Institute of Electrical and Electronic Engineers, Inc. Published 2011 by John Wiley & Sons, Inc. Vol. 1.
[24] S. Cuk, Modeling, analysis, and design of switching converters, Ph.D. thesis, California Institute of Technology, 1976.
[25] C. S. Cheng and D. Shirmohammadi, "A three-phase power flow method for real-time distribution system analysis," IEEE Transactions on Power Systems, vol. 10, no. 2, pp. 671-679, May 1995, doi: 10.1109/59.387902.

4

Load Flow Studies in a Power System

S. Mahajan, M. Sharma, V. Sood, G. Sharma

Ontario Tech University, Canada
Email: shreya.mahajan@ontariotechu.ca, manoj.sharma@ontariotechu.ca, Vijay.sood@ontariotechu.ca, gaganeep.sharma@ontariotechu.net

Abstract

The power flow study is a calculation of the power flowing in a complex, meshed power system. The study objectives are to determine the power system components, such as voltage magnitudes, voltage angles, active and reactive powers. A three-phase solution is not needed as the power flow study is a steady-state operation. If a balanced system is assumed, then a single-phase solution with a single-line diagram will be easier to model and will suffice. A power flow study is needed for the future planning and expansion of the power system. A power flow study is also used to determine the best operational mode of the power system.

Keywords: LF studies, Newton–Raphson method, computer tools

4.1 Introduction

As industrial power systems are often large and complex, they are not suitable for a hand-calculation of the power flow. In the past, dedicated analogue network analysers were used during 1930–60s to obtain physical models of the power systems. After that period, powerful digital computers became available and were used to solve for the power flow.

Digital computers were also used to perform other related studies such as stability studies (both in the transient- and steady-states), short-circuit fault calculations, and economic dispatch etc. [1]. Digital computers used were dependent on linear programming methods to obtain the optimal power flows to determine the most economic delivery of power.

A power flow assessment is essential for a complex inter-connected power system having multiple load centres. It is used for the evaluation of the power system's ability to supply power to all the interconnected loads. An assessment of total system losses, including individual line losses, is also needed. Furthermore, to ensure that the right voltage is maintained at critical locations, a determination of transformer tap settings is also needed. Thus, conducting a power flow study gives valuable insight into the system behaviour and enables the key control settings to be determined to achieve maximum capacity and reduce the operating expenses. Furthermore, computations of power flow are necessary for determining the optimal operation of generating units. The results of such power flow studies provide data on the flow of active/reactive powers, voltage magnitudes, and phase angles.

A power flow study can be split into either deterministic or uncertainty-related power flows. A deterministic power flow study ignores the uncertainties emanating from the power generations and load demands. The uncertainties can be handled in several other approaches such as probabilistic, possibilistic, information gap decision theory, and robust optimization.

4.2 Power-Flow Problem

The power flow study gathers relevant information about load active/reactive power and voltage magnitudes and angles at each bus of the power system [3]. Knowledge of this data allows the calculation of active and reactive power flows in each branch as well as the generator reactive power outputs. Since this is a nonlinear problem to be solved rapidly and with an acceptable margin, iterative numerical calculation methods have to be utilized.

The power flow study will determine that power flowing from a generator to a load via the transmission system is reliable, stable, and economically delivered. Power flow studies are essentially a prerequisite for other more elaborate power system studies. However, general-purpose digital computer programs often experience convergence problems when an interconnected radial distribution system with multiple buses is to be solved. Hence, specialized digital programs for such studies are used. Basically, a trial-and-error approach is employed, commencing with an assumed set of initial values for the total system. The power flow numbers are continuously adjusted until the result is consistent with both physical laws and meet the user-stipulated conditions. In reality, the data about the power system is provided as user input to a digital computer program and it then outputs the full details of the system. Various computational methods are used, and different kinds of data are used as input/output for the different programs. The power flow program then computes the variables pertaining to a unique, system condition. More sophisticated programs also exist i.e., optimal power flow, which looks at many hypothetical system conditions and ranks them based on user-specified criteria. Thus, the planning, design, and operation of power systems depend on power flow studies, which can be used to evaluate the effects of changes to system equipment configurations.

Traditionally, the input information is composed of:

- Bus information
- Transformer information
- Generator/motor information
- Load information

The input information is generated with the power flow output file to note the system load configuration. The following criterion about the evaluated system is then met:

- Voltage limits must not be exceeded;
- Transformers and lines power flows etc., must not exceed equipment ratings;
- Reactive power of generators must not exceed limits set by its capability curves.

The power flow study specifies a configuration with an appropriate voltage profile under normal operation. Also, the power system will be checked to see if it fails to function acceptably when certain lines misoperate due to outages, lightning strikes, or transformer failures, etc. Also, the power flow study helps to assess the reactive power supplied by capacitors for power factor correction and the settings of generator-scheduled voltages and transformer taps. The power flow study plays a vital role to determine the most ideal operation of an existing system and plan growth of the power system. In summary, a power flow study provides the real/reactive powers in every line and the voltage magnitude and phase angle of every buses.

In a typical power system, the transmission line impedance is mostly inductive; this implies that the phase angle of the transmission line impedance is nearly 90°. Hence, there is a strong link between active power and voltage angle, and between reactive power and voltage magnitude. Alternatively, this implies that the link between active power and voltage magnitude is weak. Similarly, the link between reactive power and voltage angle is weak. Consequently, active power will flow from the bus with a higher voltage angle to the bus with a lower voltage angle. Similarly, reactive power will flow from the bus with a higher voltage magnitude to the bus with a lower voltage magnitude. However, this rule-of-thumb is not true if the phase angle of the transmission line impedance is relatively small [4].

4.3 Power Flow Methods

Some of the well-known power flow solving methods that are routinely used in industry are discussed next.

4.3.1 Gauss–Seidel (GS) method

The Gauss–Seidel (GS) program was the earlier method used for power flow studies. The traditional GS method required minimal computer storage through usage of the system Y-bus matrix. Although its performance was often satisfactory on many systems, its main weakness was its long convergence time. This deficiency led to the development of Z-matrix methods. Although these methods converged more reliably, they sacrificed some of the advantages of iterative Y-bus matrix methods, such as minimal storage and high speed when applied to large systems.

4.3.2 Newton–Raphson (NR) method

The Newton–Raphson (NR) program is a more recent method used for a power flow study. It is used to solve a set of nonlinear system of equations in multiple ways. It starts with making assumptions of all the unknown variables (i.e., voltage angles at generator buses, and voltage magnitudes and angles at load buses). Next, a Taylor series expansion is used for each of the system of nonlinear power flow equations. To simply the calculation, the higher-order terms can be ignored,

4.3.3 Other power flow methods

A common extension of the NR method is the fast-decoupled power flow method. In a well-behaved power network, there is a decoupling of active/reactive flows. The fast-decoupled power flow method exploits this feature. It also fixes the value of the Jacobian matrix during the computation to avoid mathematically time-consuming matrix inversions. During the computation, it inverts Jacobian matrix only once to be efficient. To do this, it makes three assumptions. First, the conductance between the various buses is assumed to be zero. Second, the bus voltage magnitude is fixed at 1 p.u. Third, the angle between the buses is assumed zero. The fast-decoupled power flow method can obtain the answer faster than the normal NR method [5]. This is an essential need for the fast real-time solution of large power systems.

4.3.4 Load flow method using holomorphic embedding

This is a relatively new method that uses advanced techniques of complex analysis. Out of the many solutions possible in the power flow equations, this direct method rapidly obtains the solution of the operative branch.

4.3.5 Backward-forward sweep (BFS) method

This method exploits the radial structure of modern power systems. First, it chooses an initial voltage profile for the system, and then decomposes the system of equations into two separate sub-systems. The first sub-system of equations is solved using the last results of the other sub-system of equations, until a converged solution is obtained. Solving for the currents with the given voltages is called the backward sweep (BS) method [6]. Similarly, solving for the voltages with the given currents is called the forward sweep (FS) method.

4.3.6 Direct current (DC) power flow

In earlier times, DC power flows were used to emulate AC power system flows. However, DC power flow only gives an approximate estimate of line power flows in AC power systems [7]. DC power flow, due to its inherent nature, can look at active power flows only and it ignores reactive power flows. Since this method is non-iterative it converges fast, but it is less accurate than the corresponding AC power flow solution methods. DC power flow is only used wherever repetitive and fast power flow approximations are desired.

4.4 Classification of Buses

In a power system, three types of buses are defined. Each bus has four variables: voltage magnitude, voltage angle, real power, and reactive power. To begin with each bus has two known variables and two unknown variables. The various types of buses are shown in Figure 4.1.

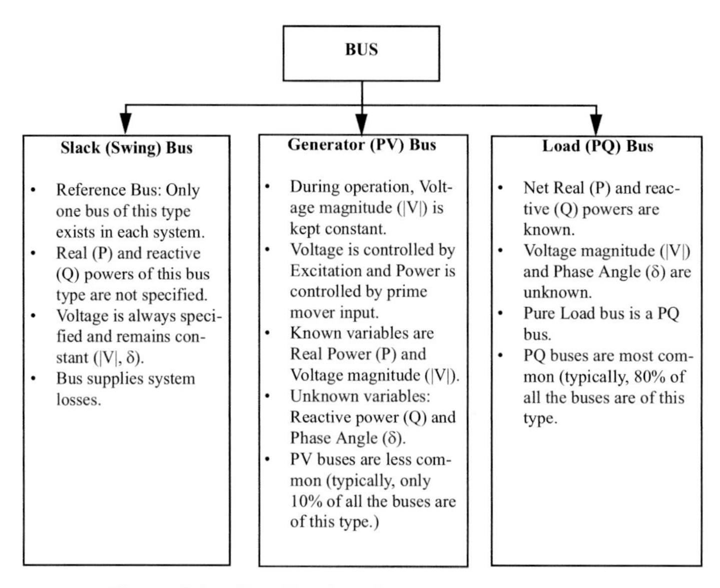

Figure 4.1 Classification of buses in a power system.

4.4.1 Reference, slack or swing bus

Traditionally, only one bus of this type is permitted in each power system. This bus is specified as the reference/slack/swing bus. The reference/swing/slack bus is different from other buses since its real and reactive powers are unspecified. It is normally tied to a generator bus with higher rating relative to the other generator buses in the power system. During system operation, the bus voltage is specified and remains fixed, both in magnitude and angle; typically, it is specified as $V_1\angle\delta_1 = 1\angle 0$ p.u. For convenience, this reference/slack/swing bus is also numbered 1. Additionally, the power losses in the power system are to be supplied by this bus. Therefore, the generator tied to the reference/swing/slack bus supplies the difference between the real power into the system supplied by the other buses and the total system output plus losses. Also, net power flows cannot be specified in advance at every generating bus as the system power losses are unknown until the study is completed. The power flow program finally calculates the real power P and reactive power Q at the reference/swing/slack bus.

4.4.2 Voltage controlled or generator bus (PV bus)

The voltage magnitude at this bus is kept fixed during the power flow calculation. A generator is always connected to this bus; the voltage is fixed by using the machine excitation and the power is controlled by using the mechanical prime mover setting. Thus, the two known variables of this bus are voltage and active power. And the two unknown variables are reactive power and phase angle. Typically, in a power system only a few buses (about 10% of all the buses) are PV buses. Additionally, the voltage can be controlled by varying the injected vars into the bus by a static var device.

4.4.3 Load bus (*PQ* bus)

A load bus (i.e., with no generator at the bus) is also called a *PQ* bus. At a bus of this type, the known variables are the active and reactive powers. And the unknown variables are the voltage and phase angle. Such buses are the most common type of buses in a power system. Typically, 80% of the buses in a power system are of this type.

For conducting the power flow analysis, input variables are given for each type of bus. The two common kinds of buses are generator and load buses. At the load bus, it is assumed that the active and reactive power consumption is known. The load buses are also known as *PQ* buses in power flow analysis. At the generator buses, P and Q are also specified. However, two problems can arise. The first problem refers to balancing the *PQ* power requirements of the power system. The second problem deals with the actual control of the generators. Therefore, it is more convenient to specify the active power P for all the generators except the one generator which is at the swing/slack bus. At this generator, the bus voltage V is used as the second variable instead of the reactive power Q.

4.5 Power Flow with NR Method

In power systems, the determination of power flow is a classical engineering problem. In cases of power flow solution, the network elements are limited to the known impedances, voltage, and current sources. At each bus, the four important variables are the magnitude of voltage $|V|$, the voltage angle δ, the active power P and the reactive power Q. At each bus, only two variables will be specified while the other two will need to be calculated [14]. Although for most buses, P and Q are given, voltage magnitude $|V|$ and voltage angle δ will need to be determined but it will vary with the bus type. Although any of the four variables can be selected, a general way of classifying the buses is shown in Table 4.1.

The set of four unknown variables that produce power balance at the specified system buses are solved by the power flow algorithm. For a typical bus

Table 4.1 Power system buses classification.

Bus type	Known variable	Unknown variable
Load bus or PQ bus	P, Q	$\|V\|, \delta$
Generator bus or PV bus	$P, \|V\|$	Q, δ
Swing bus	$\|V\|, \delta$	P, Q

i, the current that is entering the bus is given in eqn (4.1). The active/reactive power balance is given by eqn (4.2) and (4.3).

$$I_i = \sum_{j=1}^{N} Y_{ij} V_j, \tag{4.1}$$

$$P_i + jQ_i = V_i I_i^*, \tag{4.2}$$

$$P_i + jQ_i = V_i I_i^* = V_i \left[\sum_{j=1}^{N} Y_{ij} V_j \right]. \tag{4.3}$$

The power balance equation is given by eqn (4.4):

$$P_i^{\text{given}} + jQ_i^{\text{given}} = P_i^{\text{comp}} + jQ_i^{\text{comp}}, \tag{4.4}$$

where

$$P_i^{\text{given}} + jQ_i^{\text{given}} = P_i^{\text{comp}} + jQ_i^{\text{comp}}. \tag{4.5}$$

Stated in another way, the power which is specified on each bus should be equal to the power flowing into the system. The generated power is taken as positive power, which is compatible with Kirchhoff's current law (KCL) equation $YV = I$. As two unknowns exist at each bus, the power flow dimension is $2 \times N$, where N indicates the number of buses. Hence, there is a need for two additional equations on each bus. These are derived from the KCL equations for any given bus i, as follows:

$$P_i^{\text{given}} + jQ_i^{\text{given}} = P_i^{conp} + jQ_i^{\text{canp}} = V_i I_i^* = V_i \left[\sum_{j=1}^{N} Y_{i,j} V_j \right]^*. \tag{4.6}$$

Eqn (4.6) can be separated into active and reactive components according to the following equations:

$$P_i^{\text{given}} = \sum_{j=1}^{N} |V_i| \, |Y_{i,j}| \, |V_j| \cos\left(\delta_i - \delta_j - \theta_{i,j}\right), \tag{4.7}$$

$$Q_i^{\text{given}} = \sum_{j=1}^{N} |V_i| \, |Y_{i,j}| \, |V_j| \sin\left(\delta_i - \delta_j - \theta_{i,j}\right), \tag{4.8}$$

where

$$V_i = |V_i| \angle \delta_i, V_j = |V_j| \angle \delta_j, \mathrm{Y}_{i,j} = |Y_{i,j}| \angle \theta_{i,j}, \tag{4.9}$$

$$P_{\text{loss}} = \sum_{j=1}^{N} I_i^2 R_{ij} \tag{4.10}$$

where
$N =$ Total number of system buses
$|V_i| =$ Voltage magnitude at the bus i
$|V_j| =$ Voltage magnitude at the bus j,
$\delta_i =$ Voltage angle at the bus i
$\delta_j =$ Voltage angle at the bus j
$|Y_{ij}| =$ Magnitude of line admittance between bus i and bus j
$\theta_{ij} =$ Angle of the Y_{ij} element in the bus admittance matrix
$P_i =$ Net real power injection in bus i
$Q_i =$ Net reactive power injection in bus i
$P_{loss} =$ Real losses of the system
$I_i =$ Current entering bus i
$R_{ij} =$ Line resistance between bus i and bus j.

Thus, a solution to the problem is to determine a set of bus voltages that complies with the aforementioned $2{\times}N$ equations. Solving for the power flow, the NR method is expanded first in matrix form. In mixed rectangular-polar form, it can be expressed as:

$$\begin{bmatrix} P_{1\,\text{given}} - P_{1\text{comp}}^{(m)} \\ P_{2\,\text{given}} - P_{2comp}^{(m)} \\ \vdots \\ P_{\text{Ngiven}} - P_{\text{Ncomp}}^{(m)} \\ --- \\ Q_{1\text{given}} - Q_{1\,\text{comp}}^{(m)} \\ Q_{2\,\text{given}} - Q_{2\,\text{comp}}^{(m)} \\ \vdots \\ Q_{\text{Ngiven}} - Q_{\text{Ncomp}}^{(m)} \end{bmatrix} = \begin{bmatrix} J_1 = \frac{\partial P}{\partial \delta} & J_2 = \frac{\partial P}{\partial |V|} \\ J_3 = \frac{\partial Q}{\partial \delta} & J_4 = \frac{\partial Q}{\partial |V|}, \end{bmatrix} \begin{bmatrix} (\delta_1)^{(m+1)} - (\delta_1)^{(m)} \\ (\delta_2)^{(m+1)} - (\delta_2)^{(m)} \\ \vdots \\ (\delta_N)^{(m+1)} - (\delta_N)^{(m)} \\ --- \\ |V_1|^{(m+1)} - |V_1|^{(m)} \\ |V_2|^{(m+1)} - |V_2|^{(m)} \\ \vdots \\ |V_N|^{(m+1)} - |V_N|^{(m)} \end{bmatrix}. \tag{4.11}$$

Or, in abbreviated form,

$$\begin{bmatrix} \Delta P \\ \Delta Q \end{bmatrix} = \begin{bmatrix} J_1 & J_2 \\ J_3 & J_4 \end{bmatrix} \begin{bmatrix} \Delta\delta \\ \Delta|V| \end{bmatrix}. \tag{4.12}$$

The dimension of the above problem is given as (2 N-(Number of PV buses)-2) because $|V|$ updates of PV buses are unnecessary, and $|V|$ or δ updates of the slack bus are also not necessary.

4.6 Power System Losses

In developing countries like India, as an example, the average transmission plus distribution losses is about 23% of the total generated power, although some estimates as high as 50% have also been quoted. An accurate estimation of transmission and distribution losses is therefore quite essential in many jurisdictions as the level of losses directly affects the sales and power purchase agreements and can have a significant bearing on setting the utility electricity tariff.

4.6.1 Transmission lines losses

The computation of losses in transmission lines is less complicated as compared to calculating losses in transformers or distribution systems. The basic computation of power losses is based on Kirchoff and Ohm's law. Therefore, due to a simple power transmission lines configuration, calculation of line losses requires only basic knowledge of fundamental electrical principles. However, separation of these line losses does need some more explanations. For example, transmission line losses are decomposed into three separate portions: conductor heating, electromagnetic radiation, and corona losses. Since a transmission line conductor has a finite resistance, there is an I^2R or copper power loss due to AC currents flowing through it. The conductor power loss also depends on the frequency of the current because of the so-called skin effect. In a conductor, the AC current flow in the cross-section of the conductor is nonuniformly distributed. The AC current flow is more concentrated in the outer rim of the conductor due to the skin effect. Since only a very small portion of the conductor carries that AC current, line resistance is increased and adversely affects the amount of dissipated power.

4.6.2 Distribution system losses

These are power losses incurred in transit. These distribution system losses are considered as due to the energy difference between the energy sent from the generation system and the energy that the consumers received. Typically, distribution system losses are categorized into two types: technical and nontechnical. It is vital to determine the size and causes of line losses since the lost energy is often paid for by the eventual consumer. Due to the size and complexity of the distribution system, it is impractical to have metering equipment to directly measure line losses. As energy meters do not summarize the data for the same periods, and loads vary in real time, a direct measure of actual losses is not possible. Instead, the distribution utilities rely indirectly on studies to estimate the magnitude, composition, and allocation of system losses based on an annual aggregate metering information for energy purchases and sales and use system modelling methods to justify this. Such studies are performed with modelling techniques to approximately establish the magnitude, composition, and allocation of losses.

4.7 IEEE 14-Bus System Power Flow with NR Method

As a sample case, the analysis of a benchmark IEEE 14-bus system is provided next.

4.7.1 Data for IEEE 14-bus test system

The system data for this system is taken from reference [16]. The SLD of the benchmark IEEE 14-bus system is provided in Figure 4.2. The corresponding bus and line data are given in Table 4.2 and Table 4.3 respectively. Data is based on 100 MVA case.

4.7.2 Case study: Determination of optimal location of DG unit using power flow

The methodology discussed in this chapter is to attain active power, reactive power, and losses from the benchmark IEEE 14-bus system using a power flow technique. The NR method can be used to get a solution of the power flow. The procedure to calculate this is explained next:

- Test system input line and bus data is entered.
- The base case power flow is executed to get a voltage profile of the system. The total real losses can also be obtained.
- Next, the DG is located sequentially at each bus, and a power flow is executed to obtain the voltage profile after each run and the total real losses.

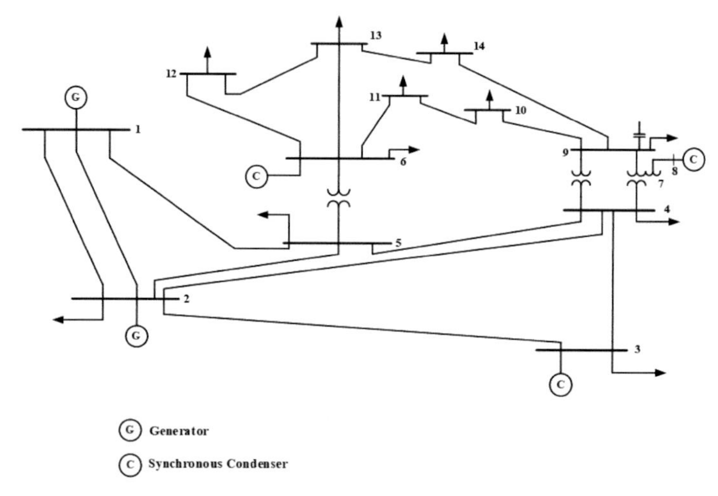

Figure 4.2 The SLD of the benchmark IEEE 14-bus system.

- The percentile improvement in the voltage profile is obtained when the DG is placed sequentially at each bus. The objective is to get any voltage difference between the value when no DG is placed, and the value

Table 4.2 Bus data for the benchmark IEEE 14-bus system (in p.u.).

Bus No.	Bus voltage		Generation powers		Load powers		Reactive power limits	
	Magnitude (p.u.)	Phase angle (Deg.)	Active power (p.u.)	Reactive power (p.u.)	Active power (p.u.)	Reactive power (p.u.)	Q_{min} (p.u.)	Q_{max} (p.u.)
1	1.060	0.000	2.324	–0.169	0.000	0.000	-	-
2	1.045	0.000	0.400	0.000	0.217	0.127	–0.40	0.50
3	1.010	0.000	0.000	0.000	0.942	0.191	0.00	0.40
4	1.000	0.000	0.000	0.000	0.478	0.039	-	-
5	1.000	0.000	0.000	0.000	0.076	0.016	-	-
6	1.070	0.000	0.000	0.000	0.112	0.075	–0.06	0.24
7	1.000	0.000	0.000	0.000	0.000	0.000	-	-
8	1.090	0.000	0.000	0.000	0.000	0.000	–0.06	0.24
9	1.000	0.000	0.000	0.000	0.295	0.166	-	-
10	1.000	0.000	0.000	0.000	0.090	0.058	-	-
11	1.000	0.000	0.000	0.000	0.035	0.018	-	-
12	1.000	0.000	0.000	0.000	0.061	0.016	-	-
13	1.000	0.000	0.000	0.000	0.135	0.058	-	-
14	1.000	0.000	0.000	0.000	0.149	0.050	-	-

Table 4.3 Line data for the benchmark IEEE 14-bus system (in p.u.).

Line no.	From bus	To bus	Line impedance (p.u.)		Half-line charging susceptance (p.u.)
			Resistance (p.u.)	Reactance (p.u.)	
1	1	2	0.01938	0.05917	0.02640
2	2	3	0.04699	0.19797	0.02190
3	2	4	0.05811	0.17632	0.01870
4	1	5	0.05403	0.22304	0.02460
5	2	5	0.05695	0.17388	0.01700
6	3	4	0.06701	0.17103	0.01730
7	4	5	0.01335	0.04211	0.00640
8	5	6	0.00000	0.25202	0
9	4	7	0.00000	0.20912	0
10	7	8	0.00000	0.17615	0
11	4	9	0.00000	0.55618	0
12	7	9	0.00000	0.110011	0
13	9	10	0.03181	0.08450	0
14	6	11	0.09498	0.19890	0
15	6	12	0.12291	0.25581	0
16	6	13	0.06615	0.13027	0
17	9	14	0.12711	0.27038	0
18	10	11	0.08205	0.19207	0
19	12	13	0.22092	0.19988	0
20	13	14	0.17093	0.34802	0

obtained after the DG is placed at a specific bus. The optimal location of the DG unit is obtained at the bus with the maximum voltage difference.

- Similarly, the percentile active power reduction is obtained. The optimum location for the DG unit is the bus number with the maximum percentile active power loss reduction.

Figure 4.3 shows the flowchart for NR-based power flow method for an optimal location of the DG unit.

4.7.3 Results and discussion

The versatility of the power flow method is demonstrated with two different benchmark IEEE bus systems.

The first test system is the IEEE 14-bus system, and the MATLAB simulation package is used to conduct the power flow. First, the voltage profile of the IEEE 14-bus system without distribution generation (DG) in service is computed. After testing, the voltage profile of this 14-bus system,

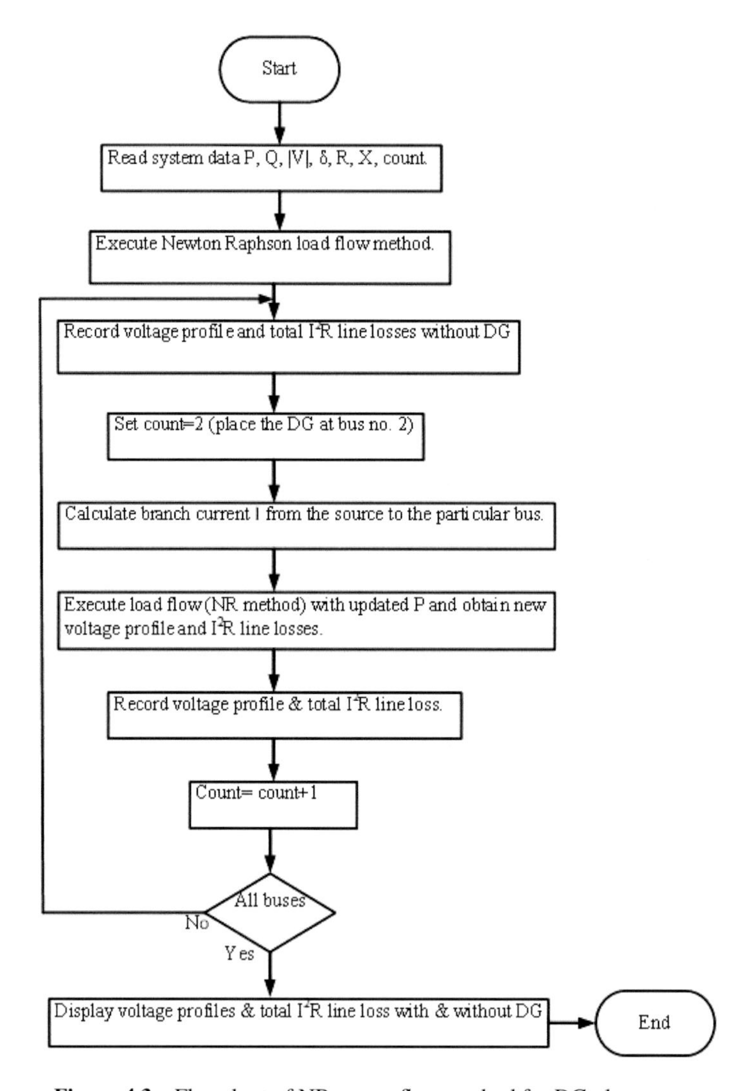

Figure 4.3 Flowchart of NR power flow method for DG placement.

with and without DG, is given in Table 4.4. The best voltage profile occurs when the DG unit is located at bus 4. The system voltage magnitudes improve significantly when the DG is located at bus 4, as shown in Figure 4.4.

Table 4.5 depicts the values of the total active power losses at the various buses after DG deployment for the IEEE 14-bus system.

Table 4.4 Voltage improvement of IEEE 14-bus system using load flow method.

Bus no.	Voltage without DG (p.u.)	Voltage with DG at bus 4 (p.u.)	% Voltage improvement
1	1.0600	1.0600	0
2	0.9761	1.0175	4.24
3	0.8924	0.9683	8.50
4	0.9130	1.0100	10.62
5	0.9270	1.0038	8.28
6	0.9165	1.0120	10.4
7	0.9005	1.0025	11.32
8	0.9005	1.0025	11.32
9	0.8850	0.9879	11.62
10	0.8917	0.9842	9.25
11	0.8947	0.9942	11.12
12	0.8978	0.9952	10.84
13	0.8911	0.9898	11.07
14	0.8661	0.9695	11.93

Table 4.5 Active power losses of IEEE 14-bus system using power flow method.

Bus location	Real power losses (MW)	Bus location	Real power losses (MW)
1	0.20852	8	0.18509
2	0.18301	9	0.17401
3	0.17366	10	0.17892
4	0.16552	11	0.18570
5	0.16772	12	0.19223
6	0.18576	13	0.18516
7	0.17562	14	0.18444

In Table 4.6, the minimum real power losses and loss reduction as a percentage for the IEEE 14-bus system are provided along with the optimum location.

Table 4.6 Optimal location of DG in IEEE 14-bus system using power flow method.

Item	IEEE 14-bus system
Best voltage profile (%)	DG located at bus 4
Real power loss without DG	0.20852 MW
Minimum power losses	0.16552 MW
Minimum power losses	25.62 MW

The voltage profile of IEEE 14-bus system is shown in Figure 4.4.

Figure 4.5 shows the line graph of the active power losses of the system versus the bus location when the DG is located at each bus of IEEE 14-bus system.

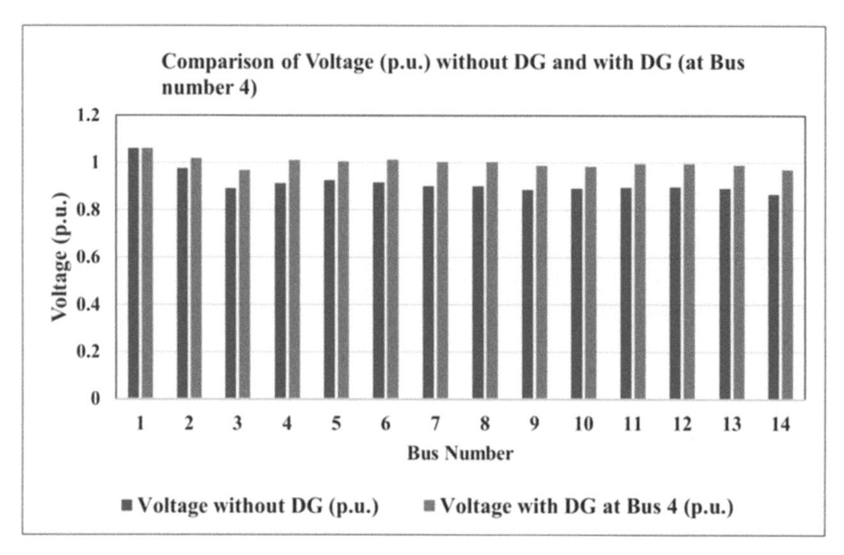

Figure 4.4 Voltage profile, with and without DG, for IEEE 14 bus system using power flow method.

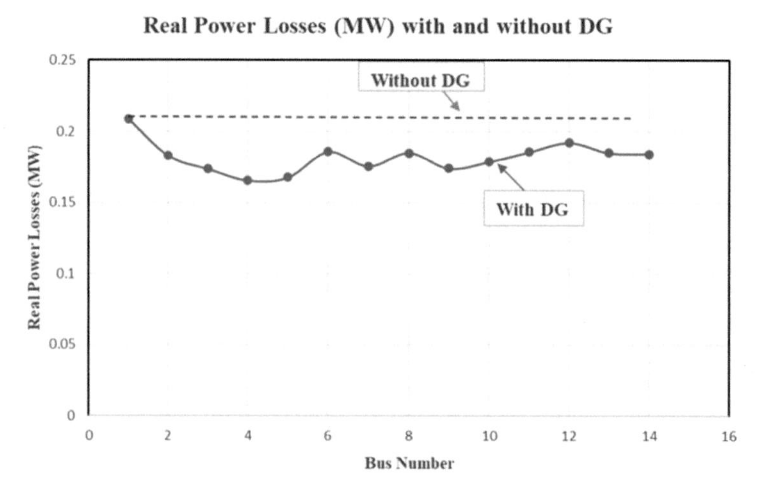

Figure 4.5 Real losses, with and without DG, for IEEE 14-bus system using load flow method.

4.8 Simulink for Power Flow Analysis

MATLAB-Simulink with its graphical user interface (GUI) environment is a popular software package for simulating many types of dynamical systems. In the Simulink package, system models can be drawn on the screen as block diagrams. A standard list of components such as connectors, loads, sources, linear and nonlinear components are included in the block library. To build a system model, the required components for the model are selected from the library. They are then assembled in the Simulink window using a mouse.

To demonstrate the flexibility of the software packages, a sample study using MATLAB/Simulink is presented next for the benchmark 5-bus IEEE System [9] is shown using MATLAB (R2022a), Simulink (Version 10.5), and Simscape (Version 5.3). The data presented on MathWorks website is used for the purpose of this analysis.

4.8.1 IEEE 5-bus model

The second benchmark IEEE bus systems chosen to showcase the versatility of the power flow is the IEEE 5-bus model. Figure 4.6 shows a single-line diagram (SLD) for this system.

The corresponding bus data for the IEEE 5-bus system is shown in Table 4.7.

Table 4.8 shows the line data for the IEEE 5-bus system.

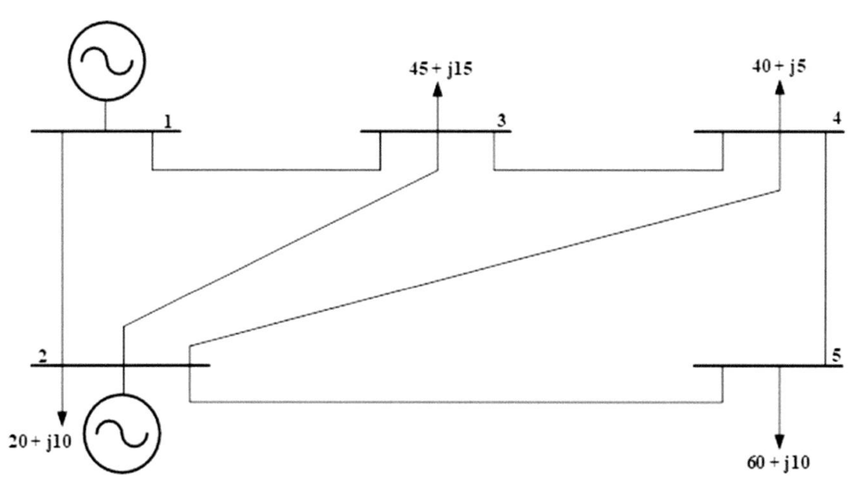

Figure 4.6 SLD for the IEEE 5-bus system [9].

Table 4.7 Bus data for IEEE 5-bus system.

Bus No	Bus Voltage	Generation		Load	
		MW	MVar	MW	MVar
1	$1.06 + j0.0$	0	0	0	0
2	$1.0 + j0.0$	40	30	20	10
3	$1.0 + j0.0$	0	0	45	15
4	$1.0 + j0.0$	0	0	40	5
5	$1.0 + j0.0$	0	0	60	10

Table 4.8 Line data for IEEE 5-bus system.

Line	Line Impedance		Line Charging
	R per unit	X per unit	
1–2	0.02	0.06	$0.0 + j0.03$
1–3	0.08	0.24	$0.0 + j0.025$
2–3	0.06	0.25	$0.0 + j0.02$
2–4	0.06	0.18	$0.0 + j0.02$
2–5	0.04	0.12	$0.0 + j0.015$
3–4	0.01	0.03	$0.0 + j0.01$
4–5	0.08	0.24	$0.0 + j0.025$

4.8.2 Analysis of 5-bus model using Simulink [9]

Using the data, the model of the IEEE 5-bus system [9] is developed in Simulink (as shown in Figure 4.7).

The base voltages of the bus (nominal rms voltages, phase-to-phase values) are also indicated in the load flow blocks. The voltages at the PV buses, or the voltage and angle of the reference/swing/slack buses are also indicated. When the power flow is solved, the bus voltage (positive-sequence value) and phase angle are displayed.

The bus type (i.e., whether PV, PQ, or Swing) is indicated by the power flow blocks tied to that bus. If there are several power flow blocks with different types (specified either in the load or generator type parameters) connected to the same bus, the power flow program automatically decides the resulting bus type (i.e., PV, PQ, or Swing).

The *powergui* block uses one of the following solution methods for solving the circuit:

- Continuous solution, which uses variable time-steps;
- Discrete solution, which uses fixed time-steps; and
- Continuous solution or discrete phasor solution.

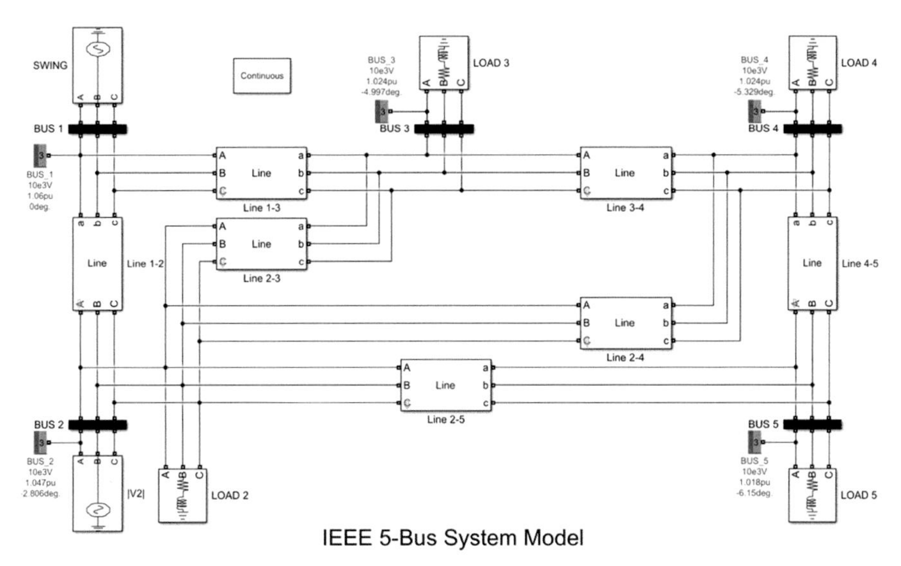

Figure 4.7 IEEE 5-bus system Simulink model [9].

The *powergui* block contains tools for steady- and transient-state analysis. This block can also be used to simulate any Simulink model which has Simscape^{TM} Electrical^{TM} Specialized Power Systems blocks. It is quite powerful and has an equivalent Simulink circuit that can represent the state-space equations of the model. When using the *powergui* block in a Simulink model, certain precautions must be taken:

- The *powergui* block is to be placed in the hierarchical top-level window in the folder.
- The block should also be identified as *powergui*.

To start the analysis, click on the *powergui* block and Block Parameters: *powergui* will pop up, as shown in Figure 4.8.

As shown in Figure 4.8, in the *powergui* menu, select the Tools tab and click the Load Flow Analyzer button.

The Load Flow Analyzer employs the *powerloadflow* function. The NR method is used for a robust and fast convergence solution. Two types of load flows are possible with the Load Flow Analyzer:

- For a balanced three-phase system, positive-sequence component power flow is obtained. Positive-sequence voltages, and PQ flows are calculated at every bus,

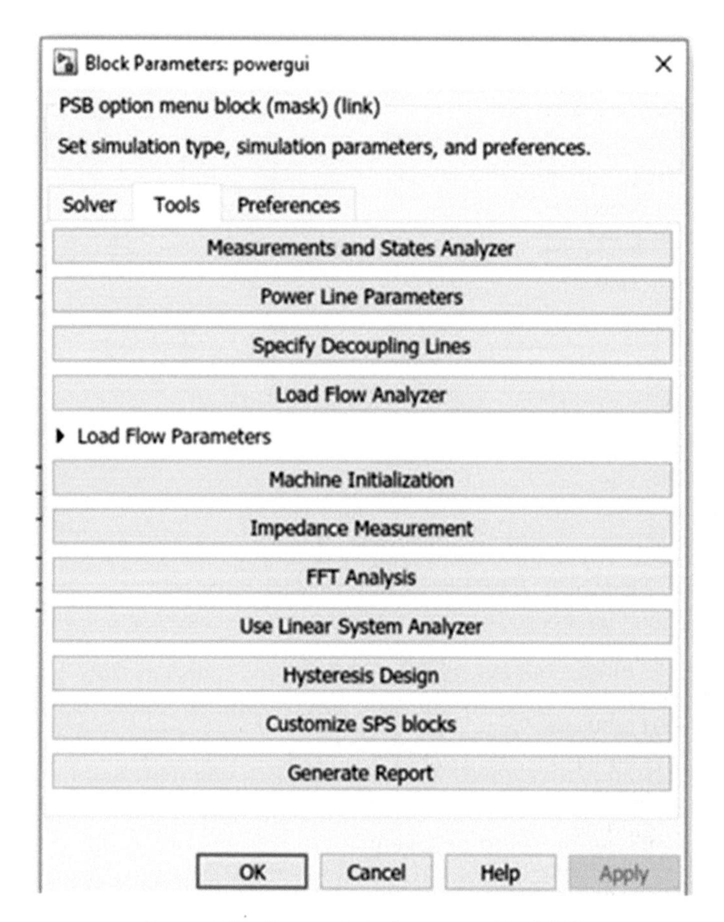

Figure 4.8 Screen shot of *powergui* block [9].

- For an unbalanced load flow, with a system having a mixture of three-phase, two-phase, and single-phase buses, individual phase voltages and PQ flows are to be computed for each bus.

To solve for a power flow, the following four quantities at each bus are to be determined Figure 4.9:

- Net active power (P) absorbed at the bus, and reactive power (Q) injected into the bus;
- Voltage magnitude (V) and voltage angle (V_{angle}) of bus voltage (positive-sequence component) or phase voltage.

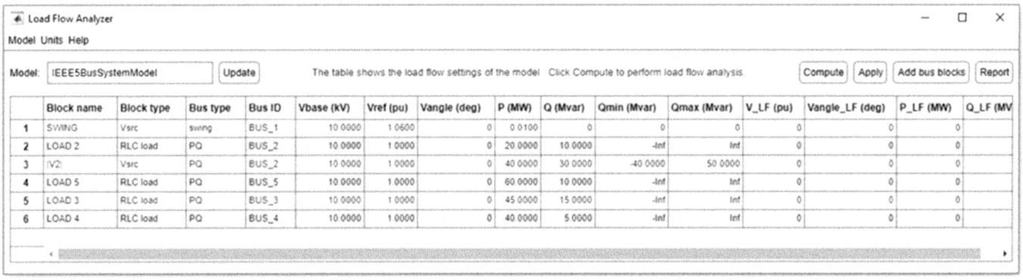

Load Flow Analyzer

Model Units Help

Model: IEEE5BusSystemModel | Update | The table shows the load flow settings of the model. Click Compute to perform load flow analysis. | Compute | Apply | Add bus blocks | Report

	Block name	Block type	Bus type	Bus ID	Vbase (kV)	Vref (pu)	Vangle (deg)	P (MW)	Q (Mvar)	Qmin (Mvar)	Qmax (Mvar)	V_LF (pu)	Vangle_LF (deg)	P_LF (MW)	Q_LF (MV
1	SWING	Vsrc	swing	BUS_1	10.0000	1.0600	0	0.0100	0	0	0	0	0	0	
2	LOAD 2	RLC load	PQ	BUS_2	10.0000	1.0000	0	20.0000	10.0000	-Inf	Inf	0	0	0	
3	V2	Vsrc	PQ	BUS_2	10.0000	1.0000	0	40.0000	30.0000	-40.0000	50.0000	0	0	0	
4	LOAD 5	RLC load	PQ	BUS_5	10.0000	1.0000	0	60.0000	10.0000	-Inf	Inf	0	0	0	
5	LOAD 3	RLC load	PQ	BUS_3	10.0000	1.0000	0	45.0000	15.0000	-Inf	Inf	0	0	0	
6	LOAD 4	RLC load	PQ	BUS_4	10.0000	1.0000	0	40.0000	5.0000	-Inf	Inf	0	0	0	

Figure 4.9 Summary of the power flow data of the model [9].

To solve for the power flow at each bus, only two out of the four quantities are known, and the remaining two quantities are unknown. Also, all the buses in the power system are classified as one of the following types:

1. *PV* bus – At a bus of this type, the active power P and generated voltage V values are specified. The bus is considered a generation bus having either a three-phase voltage source or synchronous machine. The active power P is generated, while the generator voltage V is forced. The power flow solution then calculates the machine reactive power Q that is needed to maintain the reference voltage magnitude V, and the reference voltage angle V_{angle}.
2. *PQ* bus – At a bus of this type, the specified active power P and reactive power Q are either injected into the bus (generation *PQ* bus) or absorbed by a load connected at that bus. The load flow solution computes the bus voltage magnitude V and the voltage angle V_{angle}.
3. Swing bus – At a bus of this type, the voltage magnitude V and angle V_{angle} are imposed. The power flow solution provides the active power P and reactive power Q that are generated/absorbed at that bus to balance the generated power, loads, and losses. In the model, at least one bus must be designated as a reference/swing/slack bus for a positive-sequence component power flow. It is standard practice to select one voltage source or synchronous machine as a reference/swing bus. For an unbalanced power flow calculation, it is possible to select the three/two/one phases of a 3-phase, 2-phase, or 1-phase voltage source block of an AC voltage source block as the swing buses.

Once the data is entered in the Load Flow Bus blocks, pushing the button of the Load Flow Analyzer of the *powergui* block displays a summary of the model power flow data (Figure 4.10).

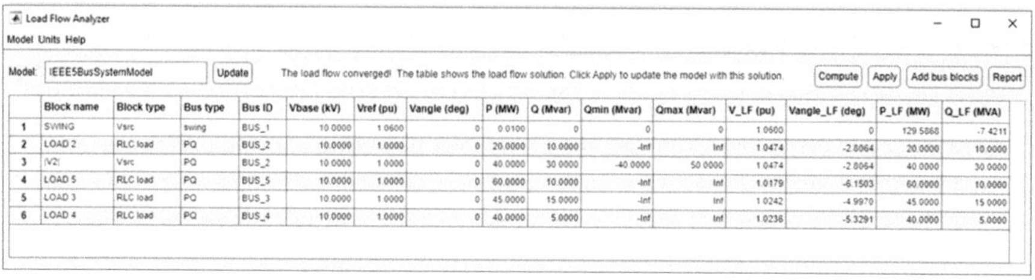

Load Flow Analyzer
Model Units Help
Model: IEEE5BusSystemModel Update The load flow converged! The table shows the load flow solution. Click Apply to update the model with this solution. Compute Apply Add bus blocks Report

	Block name	Block type	Bus type	Bus ID	Vbase (kV)	Vref (pu)	Vangle (deg)	P (MW)	Q (Mvar)	Qmin (Mvar)	Qmax (Mvar)	V_LF (pu)	Vangle_LF (deg)	P_LF (MW)	Q_LF (MVA)
1	SWING	Vsrc	swing	BUS_1	10.0000	1.0600	0	0.0100	0	0	0	1.0600	0	129.5868	-7.4211
2	LOAD 2	RLC load	PQ	BUS_2	10.0000	1.0000	0	20.0000	10.0000	-Inf	Inf	1.0474	-2.8064	20.0000	10.0000
3	|V2|	Vsrc	PQ	BUS_2	10.0000	1.0000	0	40.0000	30.0000	-40.0000	50.0000	1.0474	-2.8064	40.0000	30.0000
4	LOAD 5	RLC load	PQ	BUS_5	10.0000	1.0000	0	60.0000	10.0000	-Inf	Inf	1.0179	-6.1503	60.0000	10.0000
5	LOAD 3	RLC load	PQ	BUS_3	10.0000	1.0000	0	45.0000	15.0000	-Inf	Inf	1.0242	-4.9970	45.0000	15.0000
6	LOAD 4	RLC load	PQ	BUS_4	10.0000	1.0000	0	40.0000	5.0000	-Inf	Inf	1.0236	-5.3291	40.0000	5.0000

Figure 4.10 Computed parameters from the power flow analysis [9].

To perform power flow analysis, click "Compute" in the Load Flow Analyzer window. This will show the computed parameters as shown in Figure 4.11.

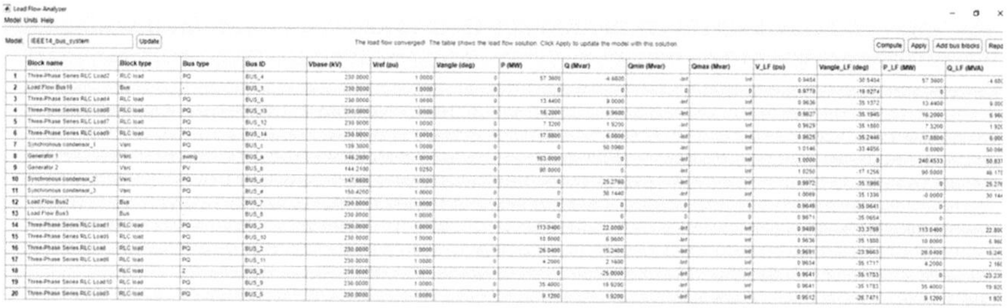

Figure 4.11 Load flow analysis [9].

On completion of load flow analysis, click "Apply" to display the result onto the IEEE 5-bus system model.

The Simulink power flow analysis result can be verified with the solution provided in the IEEE 5-bus system (Table 4.9).

Table 4.9 IEEE 5-bus system load flow solution.

Bus no.	Generation		Load demand		Bus voltage	
	MW	MVar	MW	MVar	Voltage (p.u.)	Angle
1.	129.59	–7.42	0	0	1.06	0
2.	40	30	20	10	1.0474	–2.8063
3.	0	0	45	15	1.0242	–4.887
4.	0	0	40	5	1.0236	–5.3291
5.	0	0	60	10	1.0179	–6.1503

4.8.2.1 Advantages of Simulink power load flow analysis

- One of the main advantages of using Simulink blocks is that they are user-friendly and easy to understand and manipulate.

- Simulink provides a highly interactive, easy-to-use graphic environment and a vast set of prebuilt library blocks to make modelling easy.
- We can drag and drop blocks from various libraries, configure their parameters, and connect them with wires to form the system.
- We can also create custom blocks using MATLAB functions or other Simulink models.
- The design validation is also quite easy to achieve.
- Additionally, Simulink also provides an option to create, modify, and simulate the MATLAB environment models programmatically.

4.8.2.2 Limitations of Simulink for power flow analysis

- The breakpoints cannot be set on Simscape blocks.
- The Simulink Fixed-Point Tool cannot be used with Simscape blocks.

A comparison of Simulink approach with other software i.e., Power World is shown in Table 4.10. It can be seen that Power World Simulator is much more user-friendly and permits visualization of the power system using full-colour, animated one-line diagrams, complete with zooming features.

Table 4.10 Comparison of Simulink approach with other software i.e., Power World.

S. No.	Other models (Power World Simulator)	Simulink model
1.	Power World Simulator permits visualization of the system using full-colour, animated one-line diagrams, complete with zooming features.	The visualization is neither that informative nor animated.
2.	In power flow analysis, every line has a value mentioned.	The power flow data is not directly visible on the model.

4.9 Conclusion

A power flow study provides a valuable insight into the power system operation and the power system's ability to supply power to connected loads. A calculation of the total system losses as well as individual line losses is also made. The study also provides the transformer tap settings to ensure that the correct voltage exists at critical locations. The study optimizes the generator control settings to obtain maximum power capacity from the system. The results of power flow analysis include the four variables: active power P, reactive power Q, voltage magnitude V, and phase angle V_{angle} at each bus of the system.

Due to power system complexity, a load flow study is done by either dedicated programs or general-purpose simulation programs. For both approaches, the NR approach is used. In this chapter, first a dedicated program is used to showcase the study of an IEEE 14-bus system. Secondly, a simulation study based on a MATLAB program is employed to study a benchmark IEEE 5-bus power system.

References

[1] Low, S. H. (2013). "Convex relaxation of optimal power flow: A tutorial," 2013 IREP Symposium Bulk Power System Dynamics and Control - IX Optimization, Security and Control of the Emerging Power Grid. pp. 1–06. doi:10.1109/IREP.2013.6629391. ISBN 978-1-4799-0199-9. S2CID 14195805.

[2] Aien, Morteza; Hajebrahimi, Ali; Fotuhi-Firuzabad, Mahmud (2016). "A comprehensive review on uncertainty modelling techniques in power system studies," Renewable and Sustainable Energy Reviews. 57: 1077–1089. doi:10.1016/j.rser.2015.12.070.

[3] Grainger, J.; Stevenson, W. (1994). Power System Analysis. New York: McGraw–Hill. ISBN 0-07-061293-5.

[4] Andersson, G: Lectures on Modelling and Analysis of Electric Power Systems Archived 2017-02-15 at the Wayback Machine.

[5] Stott, B.; Alsac, O. (May 1974). "Fast Decoupled Load Flow," IEEE Transactions on Power Apparatus and Systems. PAS-93 (3): 859–869. doi:10.1109/tpas.1974.293985. ISSN 0018-9510.

[6] Petridis, S.; Blanas, O.; Rakopoulos, D.; Stergiopoulos, F.; Nikolopoulos, N.; Voutetakis, S., "An Efficient Backward/Forward Sweep Algorithm for Power Flow Analysis through a Novel Tree-Like Structure for Unbalanced Distribution Networks," Energies 2021, 14, 897. https://doi.org/10.3390/en14040897, https://www.mdpi.com/1996-1073/14/4/897

[7] N. Nwulu and S.L. Gbadamosi, "Optimal Operation and Control of Power Systems Using an Algebraic Modelling Language", Springer Book, 2021.

[8] W. Tinney and C. Hart, "Power Flow Solution by Newton's Method," IEEE Transactions on Power Apparatus and Systems, vol. PAS-86, no. 11, pp. 1449-1460, 1967.

[9] Rodney Tan (2023). IEEE 5-Bus System Model (https://www.mathworks.com/matlabcentral/fileexchange/66555-ieee-5-bus-system-model), MATLAB Central File Exchange. Retrieved September 21, 2023.

5

Novel Hybrid Swarm Intelligence and Cuckoo Search Based Microgrid EMS for Optimal Energy Scheduling

Priyadarshini Balasubramanyam, Vijay K. Sood

Faculty of Engineering and Applied Science, Ontario Tech University, Canada
Email: Priyadarshini.balasubramanyam1@ontariotechu.net, Vijay.Sood@ontariotechu.ca

Abstract

A grid-connected or islanded microgrid made up of distributed energy sources (DERs), requires a power management/dispatch system to control the power dispatch and meet the load demand in the system. At the tertiary control level in a typical microgrid, an optimal scheduling mechanism is used to manage the power generated from the local DERs, energy drawn from the grid and energy consumption by the load. This chapter proposes a novel hybrid optimization technique for day-ahead scheduling in a smart-grid. A hybrid feedback PSO-MCS algorithm is implemented using swarm intelligence and cuckoo search to enhance the performance and obtain a cost-effective solution for a microgrid prosumer. A comparison has been made of the hybrid feedback PSO-MCS (HFPSOMCS) algorithm with PSO and modified CS (MCS) algorithm. The best performing algorithm among the three is executed in MATLAB/Simulink and Python IDE platforms to compare the execution time.

Keywords: Microgrid, PSO, CSA, hybrid algorithm, optimization, EMS, tertiary control

5.1 Introduction

Microgrids, an important component of modern power systems, have been rapidly evolving over the past two decades, with the goal of reaching net zero in the coming decades [1]. Economic operation of the networks besides maintaining high energy production efficiency has been the major focus area. Several strategies aiming to minimize the overall microgrid costs through economic dispatch have been discussed in [2–5].

Electricity/utility sector has seen significant changes because of worldwide technology advancements. Solar PV and wind energy farms are gaining momentum because of their widespread availability and the promise to offer an environmentally beneficial solution for the long-term sustainable electricity requirements of the future. These resources fluctuate widely, necessitating efficient and responsive power generation systems. Smart grid concepts play a crucial role in these transformations. The main problem is developing a distributed generator (DG) system for remote places. Presently, 17% of the world's population lacks access to electricity, according to recent figures. These regions are often served by diesel generators, which is not only expensive but also produces a large amount of environmental pollution. In this case, a hybrid system made up of one or more renewable energy sources or the idea of a localized generation (microgrid) has the potential to supply electricity economically.

With an ever-increasing global population and correspondingly higher electricity demands, power and automation engineers/researchers are investigating the potential of renewable energy sources. The installed capacity of renewables has grown drastically over the last several years. Due to falling PV panel prices, abundance of solar energy, government issued incentives, and installation of such farms and roof top models, the usage of renewables has experienced a significant uptick. It is difficult to regulate the fluctuating nature of the renewables while still incorporating it into an overall microgrid system. However, there are several advantages to doing so.

Authors in [6–8] give an overview of the literature on energy management systems (EMSs) that include renewable energy sources. An in-depth analysis of the EMS is discussed in [9] giving a detailed study of optimum energy dispatch. EMS in islanding mode is optimized using methods like tabu search [10], dynamic programming [11], particle swarm optimization (PSO) [12–13], and hybrid PSO fuzzy logic [14]. Astitva et al. provided a novel EMS for a microgrid with extensive focus on PV penetration and battery energy storage systems in grid connected mode of operation. A modified grey wolf optimization technique was used for microgrid control and to facilitate the economic dispatch of DERs [15].

EMS in a microgrid participates in optimization, planning, and scheduling using several control strategies, programming methods, and a set of constraints. Its performance can be enhanced by using machine learning, deep learning, data science, and data mining techniques/applications. A comprehensive and extensive review of the above-mentioned aspects of microgrids including advantages and disadvantages was studied in [16]. Battery energy storage systems (BESSs) are deployed into the hybrid microgrids to reduce fluctuations in power output, maintain main grid frequency, to alleviate transients in the system, and efficient storage of excess power [17].

A microgrid performs in an efficient manner with a robust EMS. In [18], the authors implemented a novel technique based on stochastic methods. Although, the strategy could handle the cost of operation and standby reserves, there was no technical investigation into microgrid's performance. Li et al. [19] presented an autonomous nonlinear control strategy to enable smooth transitioning from grid-connected to islanded mode and vice versa. Multi-objective optimization was implemented using a metaheuristic algorithm called the ant colony optimization technique to obtain the optimal global solution. The system's goal is to reduce the amount of electricity purchased from the utility grid, hence increasing the local independence.

This study applies a novel algorithm called HFPSOMCS to a microgrid with EMS in grid-connected, islanded, and combined modes, considering a full-day load forecast profile. Taking grid tariff rates and operational constraints into account, the proposed EMS monitored demand over 24 hours a day, at the lowest operational cost for the DGs. When demand is high or when the microgrid's generation is insufficient to meet it, power can be purchased from the utility grid. If output exceeds demand during off-peak hours, electricity can be sold back to the main grid. The MG can supply electricity to important loads even when it is detached from the main grid. Testing the case studies using Texas Instruments C2000 microcontroller to send signals to the relays of the DGs is underway.

The rest of this chapter is organized as follows. The microgrid modelling is presented in Section 5.2, and it contains schematics and mathematical functions of distinct DERs and system boundaries. The implementation of optimization algorithms to the EMS system are discussed in Section 5.3. Section 5.4 throws light on simulation results for several case studies analysed during this work. Finally, Section 5.5 concludes the chapter.

5.2 Microgrid Modelling

Figure 5.1 depicts the MG model that is utilized for the EMS and its corresponding schematic design. As part of this research, the conventional MG model is employed, which comprises several DGs, such as the combined heat and power (CHP) plant, diesel generators (DGs), and natural gas-fired generators (NGGs). A PV system, wind, and energy storage systems (ESSs) are also included in this model. At the PCC, the DGs are linked and integrated to deliver electricity to the group of loads. Operationally, the MG may be used in three distinct ways:

- **Grid-connected mode:** MG sells power back to the utility if generated power exceeds demand in this mode. MG purchases grid power when demand is greater than supply.
- **Islanded mode:** When operating in this mode, the microgrid is cut off from the utility grid and only distributes electricity to the most essential loads. During this time, a process known as "load shedding" takes place to balance production and demand.
- **Combined mode:** Due to a scheduled power outage, the system is running in grid-connected mode on a certain day transition to islanded mode. As a result, the demand for power drops off during these hours, allowing the EMS to determine the most cost-effective method of dispatching electricity to critical loads.

Table 5.1 Power ratings of the generators [28].

Lower limit (MW) $\leq$ P$_{\text{Micro source}}$ $\leq$ Upper limit (MW)
$0 \leq P_{\text{CHP}} \leq 1.5$
$0 \leq P_{\text{DG}} \leq 1$
$0 \leq P_{\text{NGG}} \leq 1$
$0 \leq P_{\text{PV}} \leq 0.5$
$0 \leq P_{\text{Wind}} \leq 0.5$
$0.05 \leq P_{\text{ESS}} \leq 0.45$
$-1 \leq P_{\text{Grid}} \leq 1$

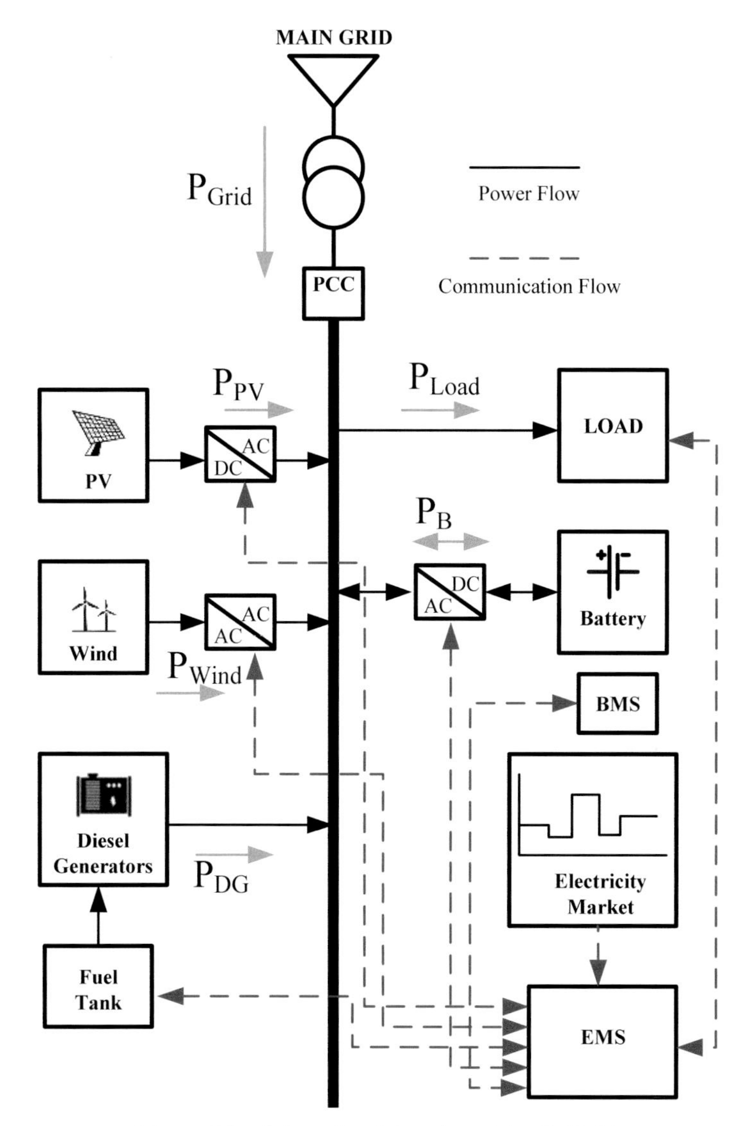

Figure 5.1 Schematic of the microgrid with EMS.

The power ratings of the system's distributed generators are shown in Table 5.1.

The mathematical modelling of each DG is formulated below as cost functions.

5.2.1 Fuel-fired generators

Fuel costs for CHP, diesel generators, and natural gas generators may all be represented mathematically using the quadratic functions shown in eqn (5.1).

$$C_{\text{Gen}}(t) = \alpha_{\text{Gen}} + \beta_{\text{Gen}} P_{\text{Gen}}(t) + \gamma_{\text{Gen}} P_{\text{Gen}}^2(t). \tag{5.1}$$

Here, α_{Gen}, β_{Gen}, and γ_{Gen} are the generator set's cost coefficients and the values considered for the study are tabulated below in Table 5.2 [26, 28].

5.2.2 PV

For a typical PV farm, the cost associated with generating and harnessing solar energy is given by

$$C_{PV}(t) = aI_P P_{PV}(t) + G^E P_{PV}(t), \tag{5.2}$$

where $P_{PV}(t)$ is the power generated by the farm in kW; I_P is the total investment cost per unit installed power ($/kW); G^E is cost of operation and maintenance ($/kW); a is coefficient of annuitization (no units) and it is calculated as follows:

$$a = \frac{r}{\left[1 - (1 + r)^{-N}\right]}, \tag{5.3}$$

where r is the rate of interest; N is investment period (= 20 years). When calculating the entire cost of generating solar energy, it is important to consider the equipment's depreciation. Eqn (5.2) and (5.3) are utilized to do this. The values for I_P and G^E are assumed to be $5000 and 1.6 cents/kW, respectively, in this case. As a result, it is possible to determine the ultimate cost function, which is shown by the eqn (5.4) [21, 28].

$$C_{PV}(t) = 545.016 \times P_{PV}(t). \tag{5.4}$$

Figure 5.2 illustrates the estimated power for this study's solar farm throughout a 24-hour period.

Table 5.2 Cost coefficients settings of various DGs.

DGs	α_{Gen}	β_{Gen}	γ_{Gen}
CHP	15.30	0.210	0.000240
Diesel generator	14.88	0.300	0.000435
Natural gas generating station	9.00	0.306	0.000315

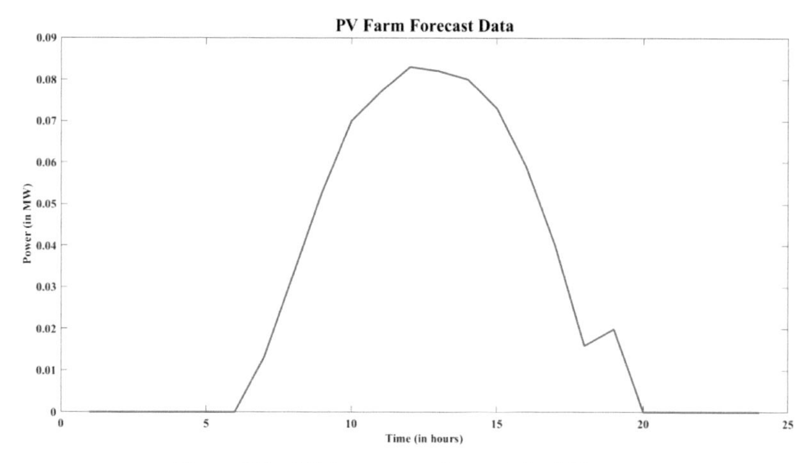

Figure 5.2 PV farm power forecast for 24 hours.

5.2.3 Wind

Wind power's cost function is comparable to that of solar generation, according to eqn (5.2) and (5.3). However, I_P and G^E are estimated to be \$1400 and 1.6 cents/kW [28], respectively. As a result, the total cost function may be determined, and it is shown as eqn (5.5) in the following way:

$$C_{\text{Wind}}(t) = 152.616 \times P_{\text{Wind}}(t). \tag{5.5}$$

For this study, Figure 5.3 shows a typical assumed generated power of the wind farm during a 24-hour period. This data is not seasonal nor regionally specific; it is generic for this experiment [22].

5.2.4 ESS

A battery with a capacity of 500 kWh was used for this investigation with a constraint that the daily number of cycles is fixed at 2. If the usage of ESS exceeds the allowed number of cycles in each day, then it is automatically disconnected from the system. The minimum and maximum SOC of ESS are set to 10% and 90% respectively, initial SOC being 50%. The battery's cost function is obtained from eqn (5.2) and (5.3), much like the two prior DGs. It is expected that the I_P and O&M expenses per unit of produced energy (G^E) are 1000 \$/kW and 1.6 cents per kilowatt hour [28]. Eqn (5.6) was used to calculate the final cost function [26].

$$C_B(t) = 119 \times P_B(t). \tag{5.6}$$

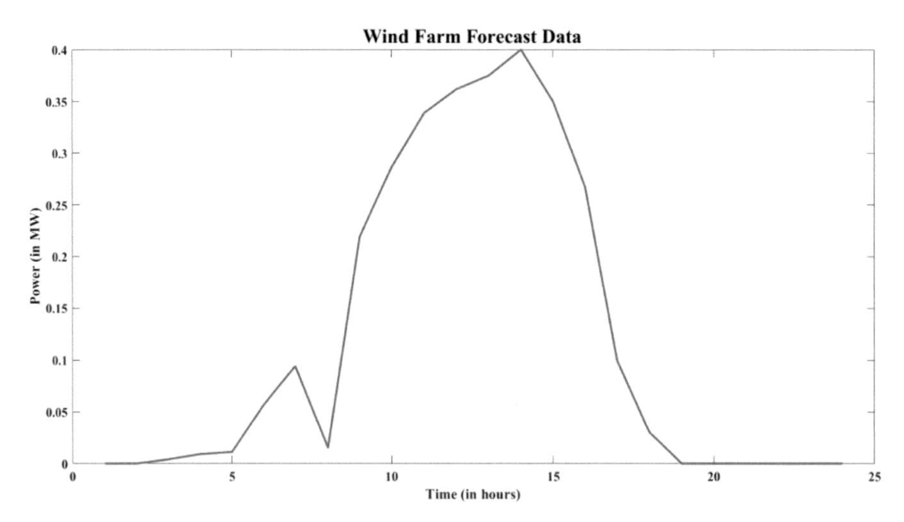

Figure 5.3 Wind farm power forecast for 24 hours.

Table 5.3 Constraints.

Mode of operation	Governing equations	Action
Grid-connected	$P_{\text{Grid}} = P_{\text{Gen}} - P_{\text{Load}}; P_{\text{Grid}} > 0$	Buy
	$P_{\text{Grid}} = P_{\text{Gen}} - P_{\text{Load}}; P_{\text{Grid}} < 0$	Sell
Islanded	$P_{\text{Gen}} = P_{\text{Load}};$	-

5.2.5 Constraints

Constraints functions are employed to aid the system in achieving its goals. Grid-connected mode allows the MG to purchase or sell power from or to the main grid based on load. If the power generated by the DGs does not match the power necessary to satisfy the load, then electricity purchased from or sold to the grid equals the difference between the power generated and the demand for the load. When the MG is in islanded mode, it is not connected to the main grid, hence there is no way to buy or sell power during this mode. Table 5.3 shows the list of various constraints included while modelling the system.

5.3 Methodology – Optimization Algorithms

This section introduces the novel optimization which is based on the outstanding behaviour of swarm intelligence and cuckoo bird search capabilities. Details of HFPSOCS algorithm and its implementation to discover the most optimal solution set to the chosen microgrid problem are presented.

5.3.1 Overview of particle swarm optimization (PSO)

PSO is one of the widely implemented meta-heuristic algorithms in various fields because of ease of implementation and faster convergence towards an optimal solution set. It is a population-based optimization algorithm discovered by Kennedy and Eberhart in 1995. This meta-heuristic algorithm

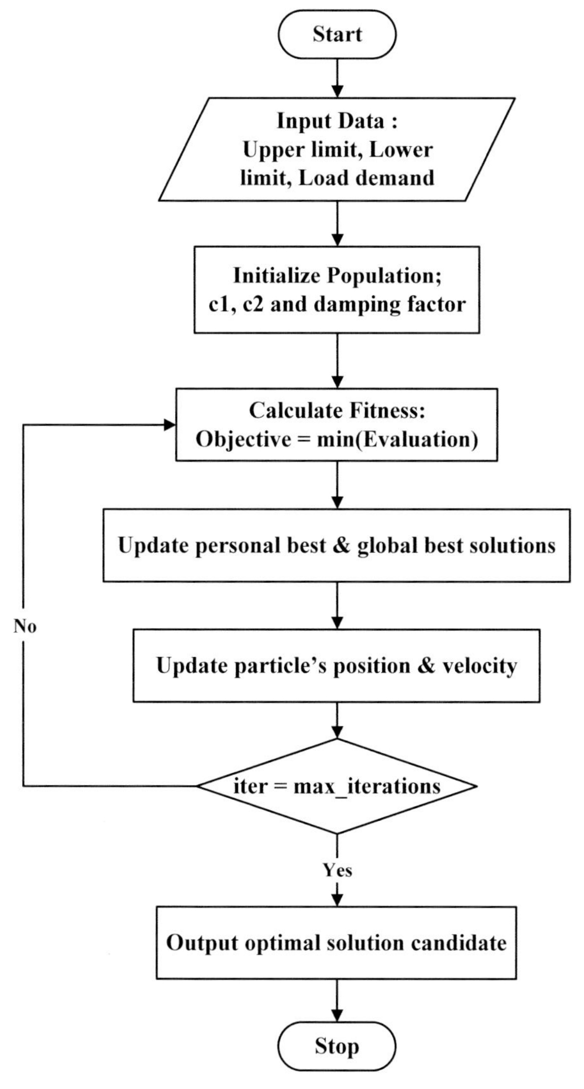

Figure 5.4 Flowchart of PSO algorithm.

Table 5.4 Pseudocode breakdown of PSO.

```
Pseudo code of the PSO algorithm:
1. Begin Process
2. Initialization
      A. Params:
            C1: Cognitive component
            C2: Social component
            W: Inertia factor
            W_damp: Damping factor
            Np: Swarm population size
      B. Generate initial swarm using randomly uniformly distributed
         function
      C. Evaluate fitness function
3. Main Loop
      for (iter = 1, iter ≤ iter_max, iter + +) Do
         for (pop = 1, pop ≤ N_p, pop + +)Do
            //Evaluate particle velocity
            v_pop(t) = w × v_pop(t − 1) +
                              c1 × r1 × (P_LBest − (t − 1)) +
                        c2 × r2 × (P_GBest − X_pop(t − 1))
            //Update particle position

               X_pop(t) = X_pop(t − 1) + v_pop(t)

               If f(X_pop) ≤ f(P_LBest) then P_LBest = X_pop
               If f(X_pop) ≤ f(P_GBest) then P_GBest = X_pop
                     Update X_pop & v_pop
         End for
      End for
4. Best Solution
5. End Process
```

uses swarm intelligence consisting of particles or swarms. Each swarm is cognate with position and velocity vectors, namely x_{ij} and y_{ij}. The dimension of the search space determines the size of these vectors. The PSO, much like other evolutionary algorithms, initializes a swarm (a set of candidate solutions) and then searches for the best possible global optimum. It considers several particles to be potential solutions and each particle moves across the search space at a certain velocity to locate the optimal candidate set. The term P_{LBest} refers to the best local solution that a particle has come up with as iterations go on. The P_{LBest} that is superior to all the other particles is referred to as the P_{GBest}. Each particle must consider its current location, its current velocity, and the distance to local best and global best before it can update its position. Each particle attempts to improve by mimicking successful peers. The best place in the search space that each particle has ever visited may also be stored in its memory [20, 23].

$v_{\text{pop}}(t)$ is used to determine the velocities of the particles after each iteration. After that, $X_{\text{pop}}(t)$ is used to determine where the particle position is updated to. Until a stopping criterion is reached, the particle position keeps changing. The flowchart (shown in Figure 5.5) is more clearly defined by the pseudo code in Table 5.5 and provides a step-by-step approach to the algorithm.

5.3.2 Overview of modified CSA

Inspired from the lifestyle and brooding behaviour of the cuckoo bird, this algorithm was introduced by Xin-She-Yang and Suash Deb in 2009 [27]. Instead of using basic isotropic random-walks, Levy flights is used to improve the performance of the technique. The standard CSA uses the following three important rules, in solving an optimization problem:

1. Each cuckoo bird picks a nest at random and lays one egg every time.
2. The highest quality eggs (solutions) in the best nests will be passed down to the future generations of birds.
3. Given that the number of host nests is invariable, the probability of discovering an egg laid by cuckoo bird in the host nest is $p_a \epsilon\ (0, 1)$; 1 being discovered. In this circumstance, the host bird has two options: either remove the egg from the nest or abandon the nest and construct an entirely new one.

For implementation, the following representation is adopted: each egg in a nest symbolizes a solution. The goal is to leverage new and superior options (cuckoo bird) to replace poorer quality solution/eggs. By controlling the switching parameter p_a, the local search random walk and global explorative capability can be improved. The equation to express local random walk is given as follows:

$$\sigma_u = \left(\frac{\Gamma(1 + \beta) \times \sin(\pi + \frac{\beta}{2})}{\Gamma\left(\frac{(1+\beta)}{2}\right) \times \beta \times (2^{\frac{\beta-1}{2}})}\right)^{1/\beta} ; \sigma_v = 1. \tag{5.7}$$

Also included are two search capabilities – local-search and global-search, which are governed by a switching probability and/or discovery probability. As mentioned in the earlier paragraph, the local-search (for p_a = 0.25) is relatively time-consuming, accounting for around one-fourth of the whole search time, whereas the global-search accounts for approximately

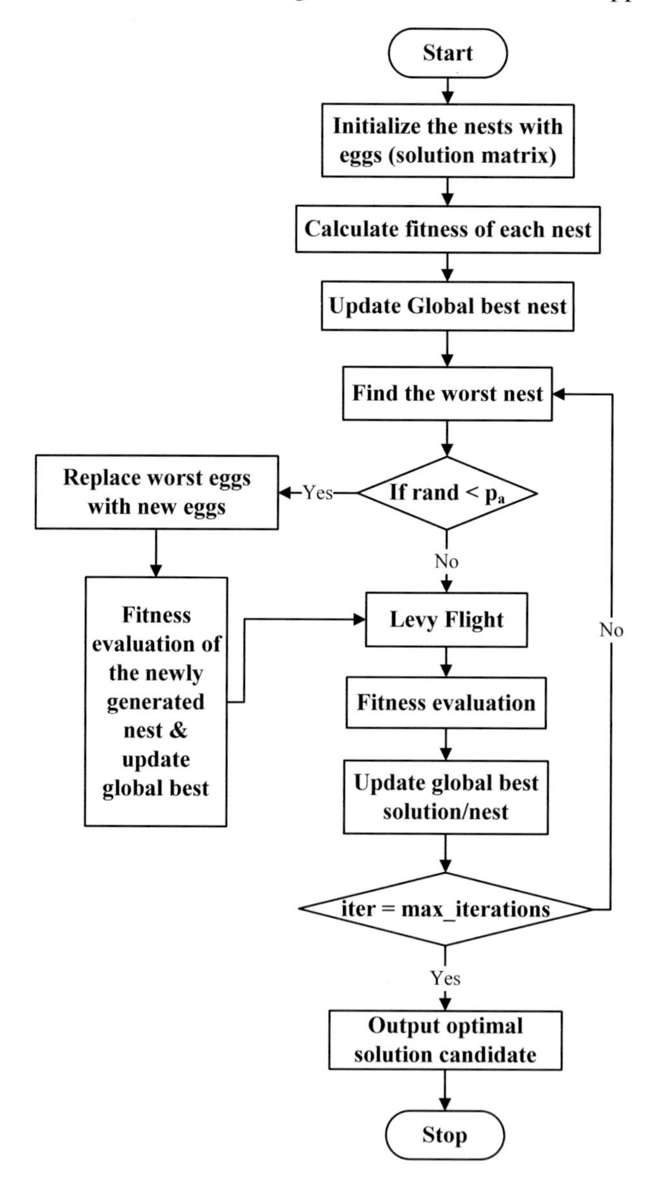

Figure 5.5 Flowchart of MCS algorithm.

three-fourths of the total search time. As a result, the global optimality can be found with a higher probability because of a more efficient global search [24].

The fact that it employs Lévy flights rather than normal random walks as part of its global search is an additional benefit of CSA. CSA can traverse the search space more efficiently than other meta-heuristic methods using the ordinary Gaussian process because Levy flights have an infinite mean and variance. The flowchart (Figure 5.5) defined in Table 5.5, gives a step-by-step approach to the algorithm.

Table 5.5 Pseudocode breakdown of MCS.

Pseudo code of the MCS algorithm:

1. Begin Process
2. Initialization
   a. Params:

      Np: Population Size

      max_iter: Maximum number of iterations

      p_a: probability of discovery rate of alien eggs

      sigma: 0.6966, beta: 3/2

      a,*b*: Two positive random numbers, $\exists\ 0 < a < b < 1$

   b. Initialize swarm and evaluate fitness and update p_g
3. Main Loop

   While (iterations <= max_iter)

   for *i* from 1 *to Np*: Do

   //Calculate Self-Adaptive step size

   $$\alpha_t = (b - a) \times \frac{(e^{\frac{10\times(t-1)}{t^M-1}} - 1)}{e^{10}-1}$$

   //Update position by applying Levy flights

   randomStep = findStep()

   $$x_i(t + 1) = x_i(t) + randomStep \times (x_i(t) - p_g)$$

   //Generate Gamma exemplar for x_i

   gammaExemplar()

   // Update global best

   If $f(x_i) <= f(p_g)$ then $p_g = x_i$

   End For

   End While
4. Best Solution
5. End Process

5.3.3 Overview of HFPSOMCS

The prime motivation behind the conceptualization and implementation of hybridized algorithm is that PSO can converge swiftly to the local best solution, though not always to the global best candidate set. According to some evaluations, it is implied that the mathematical equations used in PSO do not meet the global convergence criteria, and so there is no certainty of convergence to the global solution set. MCSA, on the other hand, has proven to satisfy the global convergence criterion and so ensures global convergence. This means that when it comes to solving multimodal optimization case studies, cuckoo search may generally converge to the optimality while PSO may converge to a local optimum set.

In contrast, MCSA takes several iterations to get the global best in the specified search space. A consequence of this is a significant increase

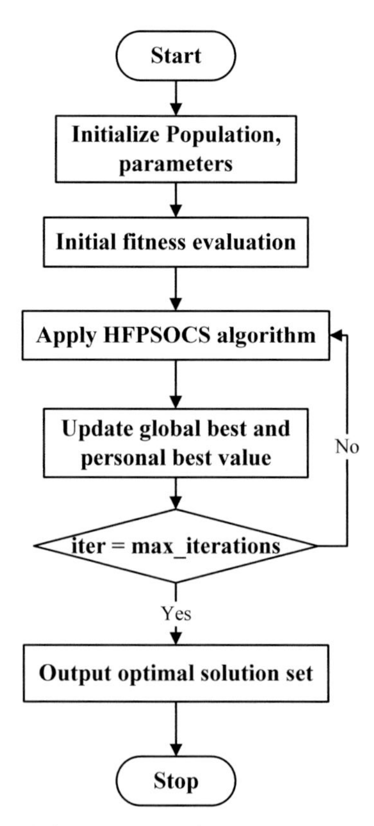

Figure 5.6 Flowchart of HFPSOMCS algorithm.

Table 5.6 Benchmark functions with constraints.

Function name	Dimension	Problem	Variable bounds	Global minimum
Sphere function	30	$f(x) = \sum_{i=1}^{d} x_i^2$	$x_i \in [-5.12, 5.12]$	$f(x^*) = 0,\ at\ x^* = (0,\ldots,0)$
Rastrigin function	30	$f(x) = 10d + \sum_{i=1}^{d} (x_i^2 - 10\cos(2\pi x_i))$	$x_i \in [-5.12, 5.12]$	$f(x^*) = 0,\ at\ x^* = (0,\ldots,0)$

in computing time, which is a disadvantage in scheduling applications. A novel feedback algorithm has been developed to achieve a balance between two desirable characteristics. Figure 5.6 illustrates the flow chart of the hybridization technique.

For the benchmark functions displayed in the table, the hybrid algorithm was used to optimize the performance of the system. Table 5.6 lists the parameters that were selected for use in the system. Traditional PSO and HFPSOMCS methods yielded the optimization results depicted in Figure 5.7. The hybrid algorithm outperformed PSO in both functions tested here. Afterwards, the hybrid system was used to resolve the EMS problem in the microgrid under consideration. After that, the findings and case studies are reviewed and developed more fully.

Figure 5.8 depicts the generic EMS architecture that is a part of the tertiary control layer.

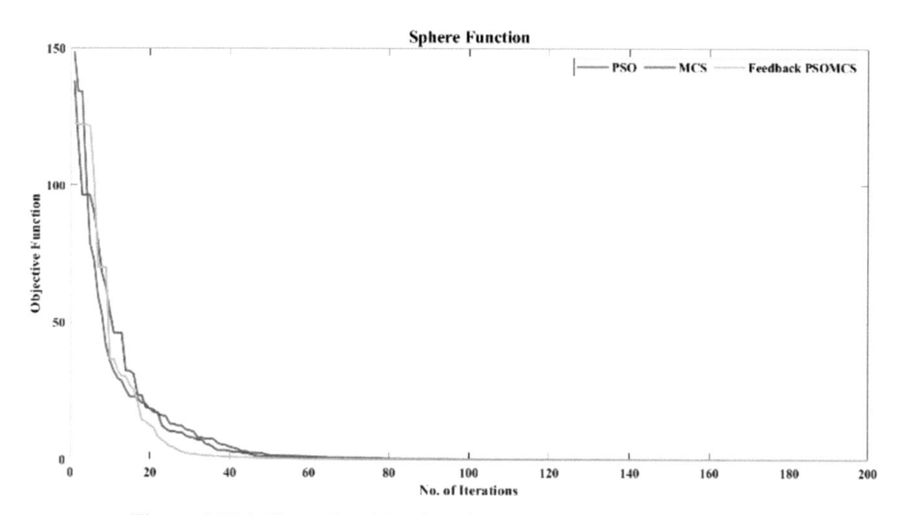

Figure 5.7(a) Three algorithms' performance on the Sphere Function

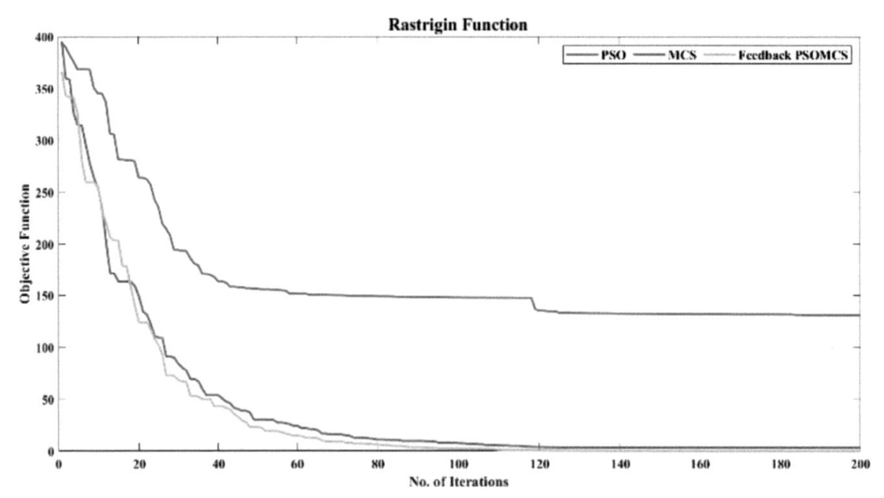

Figure 5.7(b) Three algorithms' performance on the Rastrigin Function

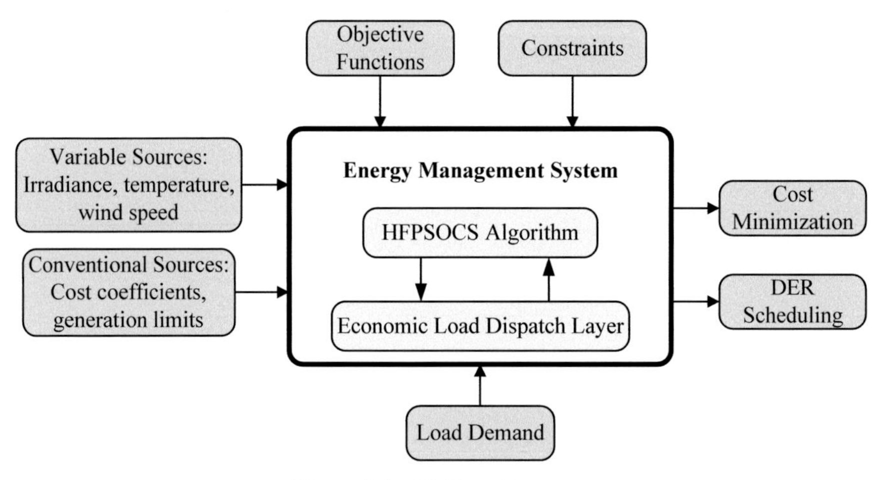

Figure 5.8 EMS architecture.

5.4 Case Studies and Results

In this section, the findings of EMS are discussed. Investigation is carried out for a full day (24 hours operation) to determine the most efficient dispatch of the DERs and utility grid. Microgrid is subjected to three different case studies, as follows:

Figure 5.9 depicts the load demand for a 24-hour time, which was utilized for all the case studies. There is a large difference in load demand between the

First Case study	Grid-connected mode	Summer season tariff rates
		Winter season tariff rates
Second case study	Islanded/Standalone mode	Summer season tariff rates
		Winter season tariff rates
Third case study	Combined mode (grid + islanded)	Summer season tariff rates
		Winter season tariff rates

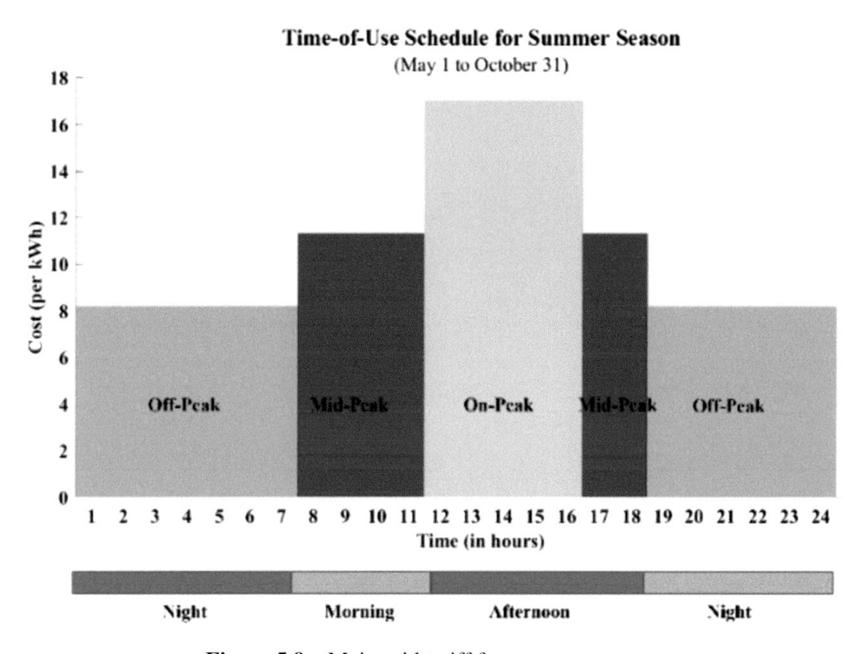

Figure 5.9 Main grid tariff for summer season.

islanded and the grid-connected modes, as seen in the Figure 5.9. As a result, load shedding is necessary, and the system only provides power to vital loads while the system enters standalone mode.

The main grid tariff is obtained from the website of the Oshawa PUC, with tariffs effective November 2021 [25]. The case studies are carried out during both the summer and winter seasons, with the primary goal of understanding BESS participation and the secondary goal of investigating the interaction between the main grid and DG units. Figures 5.9 and 5.10 show time-of-use schedule for a 24-hour operation period dependent on operating circumstances. When it comes to summer, there is a distinct midday peak hour. This is the busiest time of the year for air conditioners. Early morning and evening are two of the busiest times of day in the winter. When it is cold

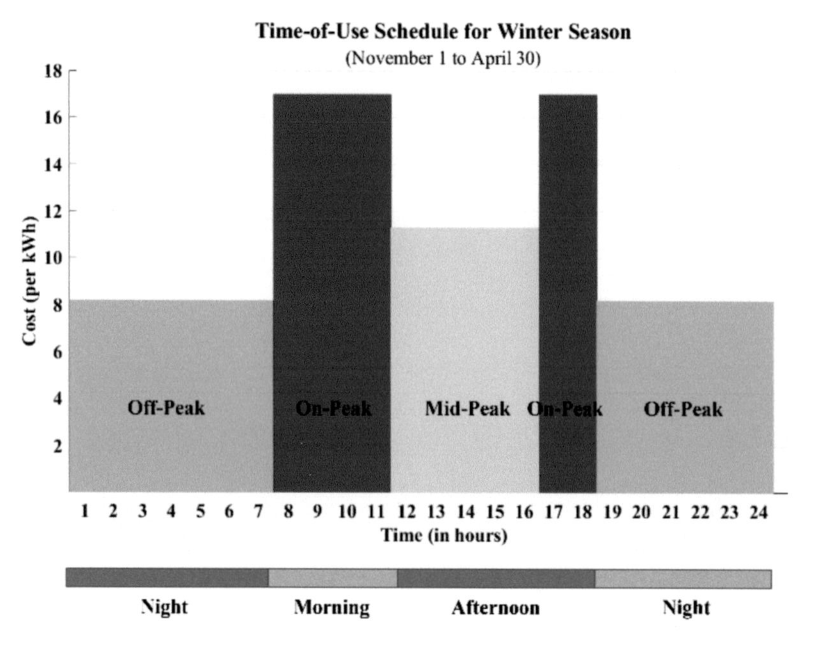

Figure 5.10 Main grid tariff for winter season.

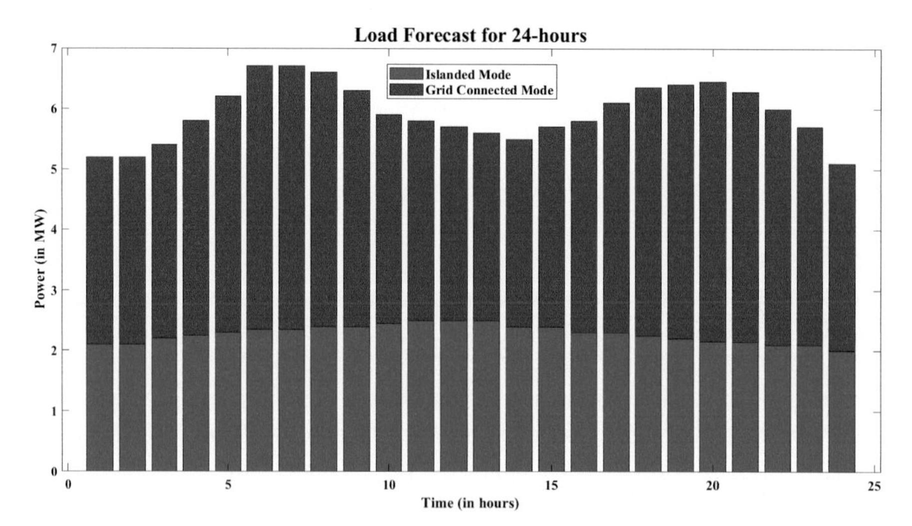

Figure 5.11 Load demand forecast for a 24-hour period.

outside, people tend to use more heating, lighting, and appliances in their homes.

5.4.1 First case study: Grid-connected mode

All the system's power needs are covered by the DG units, ESS, and the main grid in this mode.

CHP, diesel generator, natural gas generators, wind, solar, energy storage systems (ESS), and the main grid are all examined in this case study during a 24-hour operating period. Load demand, main grid tariff, and wind and PV power availability are all taken into consideration in an optimal EMS model. In addition, it is assumed that all the generators are running, and the objective of the optimization algorithm is to identify the approach that results in the lowest possible operating cost while still satisfying the demand for electricity. Table 5.7 and Figures 5.12 and 5.13 summarize the findings of this case study in both seasons.

Figures 5.12 and 5.13 show the EMS in grid-connected mode at its optimum output. The algorithm was able to identify the best way to deploy DGs to meet the given demand during the time-period. System purchases and sells electricity from the grid during both off-peak and peak periods. The total cost of operation over a day with a summer tariff is \$1690.03 whereas with a winter tariff it is \$1852.95.

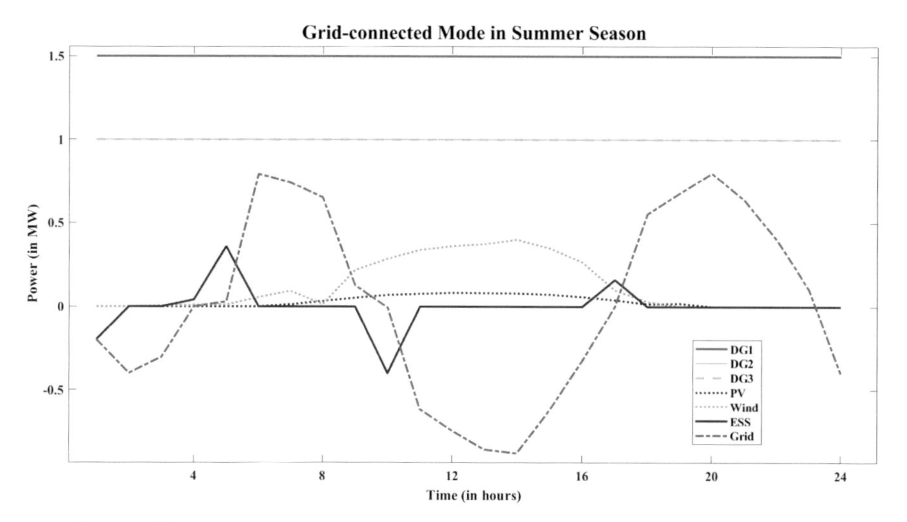

Figure 5.12 EMS optimum dispatch for grid-connected mode with summer tariff.

Table 5.7 breaks down the MG's hourly operational costs. When demand is lower than production, as shown in the table, operating costs are lower than those experienced when demand is higher [5–9], [17–23]. As a result, the MG may save money by selling excess power back to the main grid. Similarly, if output falls short of demand, the utility must pay additional fees for power from the main grid. This indicates that the algorithm can make decisions about when to purchase and/or sell power to and from the main grid, which is beneficial for the system. The convergence trend for three different algorithms

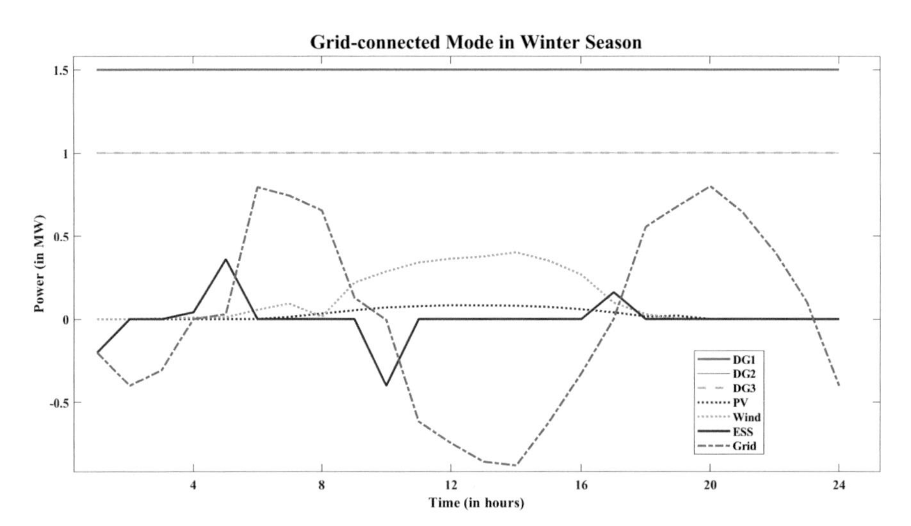

Figure 5.13 EMS optimum dispatch for grid-connected mode with winter tariff.

Table 5.7 MG's hourly operation cost breakdown (Case 1).

Hour	Cost (in $)		Hour	Cost (in $)	
	Summer	Winter		Summer	Winter
1	47.50	46.90	13	−2.54	14.59
2	7.30	6.10	14	3.65	13.94
3	15.79	14.88	15	28.44	40.90
4	46.38	46.38	16	58.38	64.90
5	86.99	87.08	17	96.46	96.46
6	113.92	116.30	18	116.09	150.99
7	122.74	124.97	19	127.84	170.68
8	134.10	175.17	20	105.70	108.10
9	117.53	125.59	21	92.58	94.50
10	169.68	169.30	22	72.90	74.10
11	65.21	26.40	23	48.30	48.60
12	15.02	29.92	24	7.30	6.10

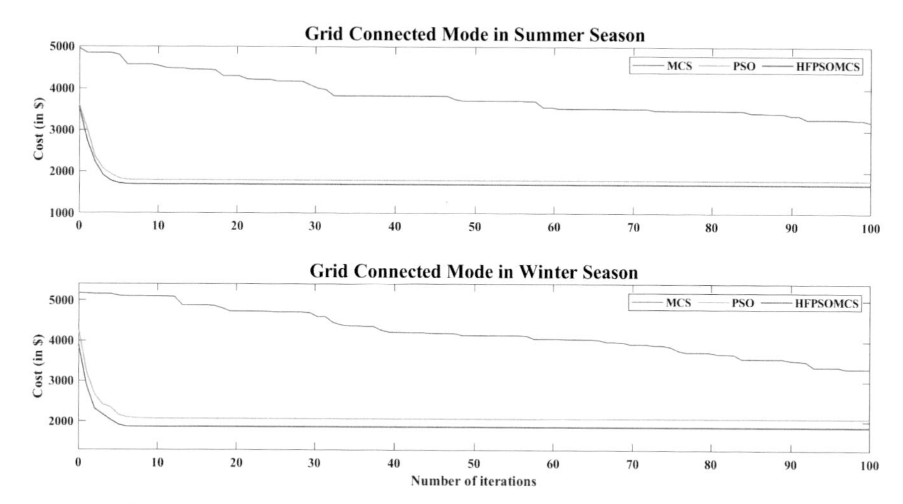

Figure 5.14 Cost convergence characteristics in grid-connected mode.

– namely, PSO, MCS, and HFPSOMCS – is depicted in a single graph in Figure 5.14. Because it can bring together the excellent search capabilities of MCS in tandem with the quicker convergence feature of PSO, the hybrid algorithm is able to converge at a faster rate than the other two optimization algorithms, as can be observed. The reason for this is because the hybrid algorithm was designed to maximize efficiency.

5.4.2 Second case study: Islanded mode

A more advanced EMS has been installed in an MG to cut down on the amount of energy needed to cater the load for a whole day, just as was done in the previous investigation. In this situation, however, the MG has been separated from the main grid and is operating in an islanded state. Figure 5.11 illustrates the load requirement that was considered appropriate for this scenario. With no choice for power trading, this study's algorithm will be forced to choose the best strategy to dispatch three available generators. PV and wind power plants are expected to be fully operational throughout each of these time periods.

The optimal output that the EMS was able to obtain for the islanded mode of operation is shown in Figures 5.15 and 5.16. The algorithm was able to determine the best way to distribute each generator so that it could meet the specified load requirement at each interval. For this scenario, the 24-hour total operating cost with summer tariff is $1788.281. With winter tariff, the

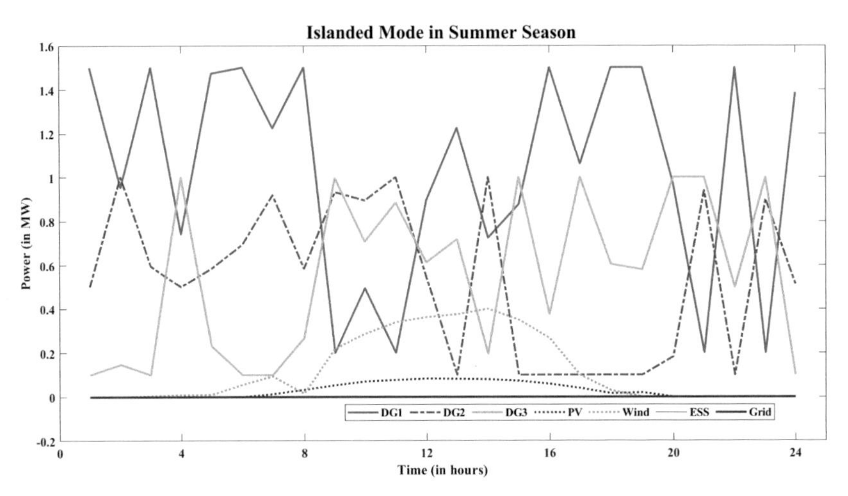

Figure 5.15 EMS optimum dispatch for islanded mode with summer tariff.

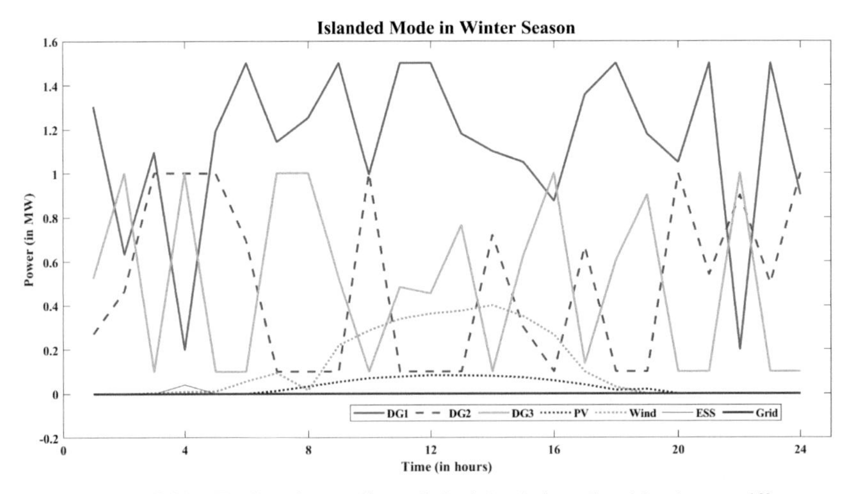

Figure 5.16 EMS optimum dispatch for islanded mode with winter tariff.

cost of operation is \$1792.92. The slight difference is because the way DGs are fired up is entirely different in both scenarios.

As can be seen in Table 5.8, the MG's operating costs when in islanded mode are broken down into discrete time intervals. In islanded mode, the system only provides power to essential loads, hence peak and off-peak hours were not significant for this case study's load demand. There is a noticeable difference in price between periods 9 and 16 hours, based on the

Table 5.8 MG's hourly operation cost breakdown (Case 2).

Hour	Cost (in $)		Hour	Cost (in $)	
	Summer	Winter		Summer	Winter
1	39.67	39.69	13	142.73	142.73
2	39.72	39.75	14	145.54	145.50
3	40.32	40.36	15	133.95	133.93
4	41.19	46.10	16	113.34	113.40
5	41.44	41.47	17	76.94	76.91
6	48.44	48.44	18	53.09	53.09
7	61.45	61.46	19	50.60	50.63
8	60.08	60.11	20	39.74	39.73
9	102.77	102.65	21	39.81	39.68
10	122.42	122.37	22	39.67	39.79
11	134.51	134.39	23	39.79	39.67
12	141.28	141.23	24	39.65	39.70

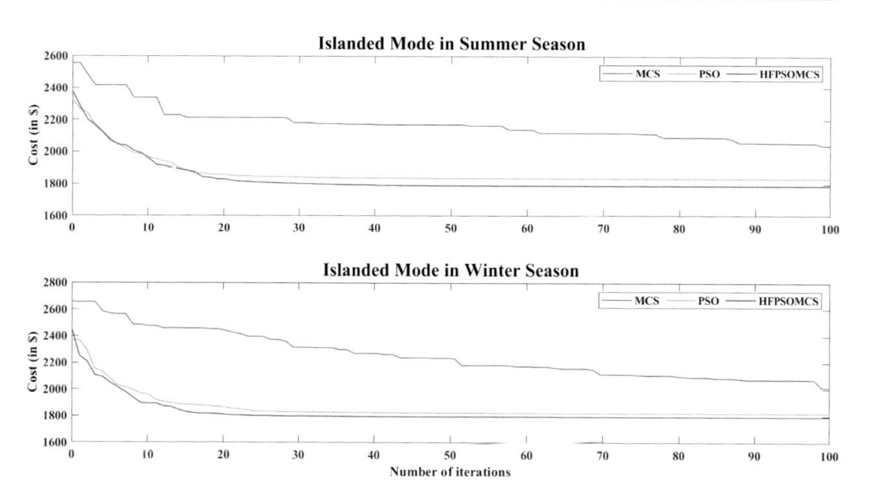

Figure 5.17 Cost convergence characteristics in islanded mode.

data presented. All three generation sources – DGs, solar photovoltaics, and wind turbines – are operating at or near their full capacity because of the high demand for electricity. Figure 5.17 depicts the convergence of operational costs in this mode.

5.4.3 Third case study: Combined mode

In this situation, the MG system that is functioning in grid-connected mode makes the transition into islanded mode for either a finite or an infinite amount of time. When it comes to everyday life, this kind of situation is not out of the question. The master controller, also known as the "brain" of the EMS, is responsible for turning off the power to the system's noncritical loads and isolating the main grid at the PCC when certain events occur, such as when equipment fails, fuses blow, scheduled maintenance is performed, natural disasters occur, and so on.

The optimum outputs for dispatch are shown in Figures 5.18 and 5.19 when the system is operating in combined mode. The load is decreased to

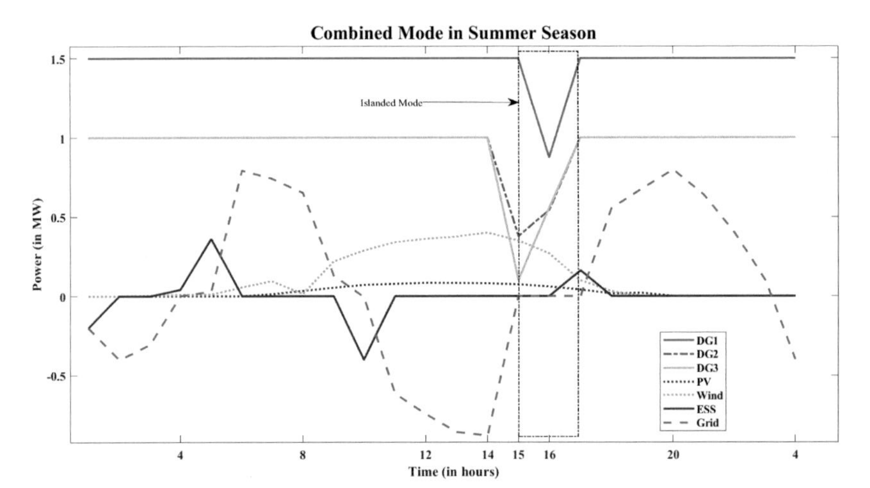

Figure 5.18 EMS optimum dispatch for combined mode with summer tariff.

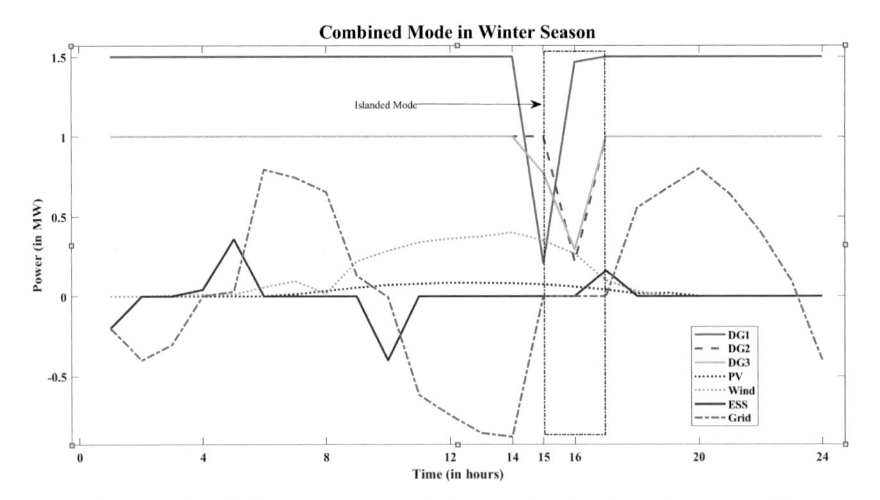

Figure 5.19 EMS optimum dispatch for combined mode with winter tariff.

50% as the system transitions from grid-connected mode to islanded mode between the 14th and 15th hours of the day. The cost of operation increases to $2005.30 when using the winter tariff, up from $1855.87 when using the summer rate. Because of the availability of power from PV and wind farms in addition to the contribution of ESS, the generators are ramped up and down when in islanded mode. However, they do not operate at their full capacity because of the contribution of ESS.

Table 5.9 MG's hourly operation cost breakdown (Case 3).

Hour	Cost (in $)		Hour	Cost (in $)	
	Summer	Winter		Summer	Winter
1	47.50	46.90	13	−2.54	14.59
2	7.30	6.10	14	3.65	13.94
3	15.79	14.88	15	133.89	134.01
4	46.38	46.38	16	113.40	113.34
5	86.99	87.08	17	96.46	96.46
6	113.92	116.30	18	116.09	150.99
7	122.74	124.97	19	127.84	170.68
8	134.10	175.17	20	105.70	108.10
9	117.53	125.59	21	92.58	94.50
10	169.68	169.30	22	72.90	74.10
11	65.21	26.40	23	48.30	48.60
12	15.02	29.92	24	7.30	6.10

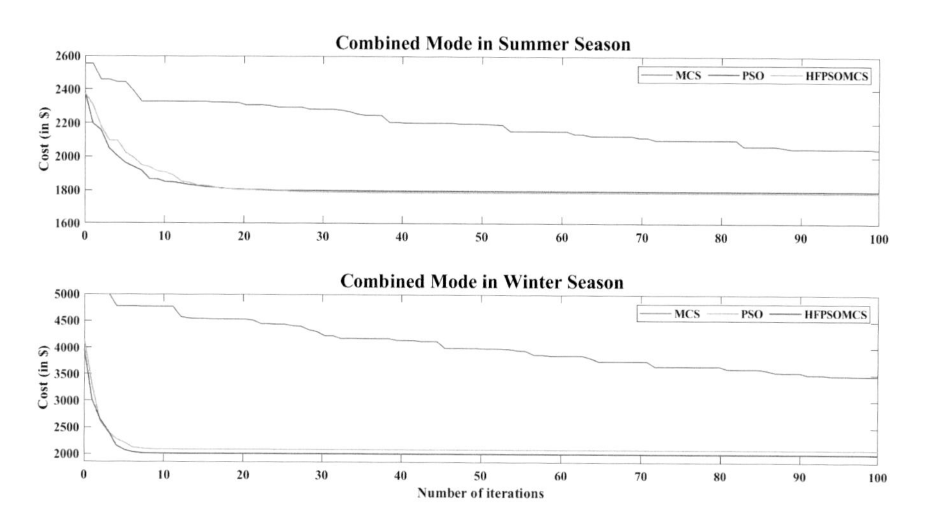

Figure 5.20 Cost convergence characteristics in combined mode.

The breakdown of the microgrid's costs, broken down by the hour, may be found in Table 5.9. Figure 5.20 depicts the convergence of operational costs in this mode.

5.5 Conclusion

This chapter deals with an EMS to regulate the power flow, balance production to the demand, to keep the grid energy costs as low as possible using an optimization algorithm in a generic microgrid.

The microgrid is composed of a CHP plant, a natural gas generator, a diesel generator, a PV energy source, a wind energy source, and a battery energy storage system, as well as critical and noncritical loads. The microgrid can operate in either grid-connected, islanded, or a combination of the first two modes. An AI-based EMS was developed using mathematical modelling of distributed energy sources and the main grid, and it can make decisions based on time-of-use and day-ahead forecasting. The cost functions are power- and time-dependent equations.

As described in Section 5.4, three algorithms, i.e., PSO, MCS, and HFPSOMCS, were implemented on three case studies for this research work. A robust feedback-based hybrid PSO-MCS optimization technique is developed with the goal of optimizing the EMS and improving battery life. By combining the best aspects of two different algorithms (PSO and MCS), the novel technique resulted in optimal scheduling and cost-saving solutions of energy management components in grid-connected, islanded, and combined modes of operation by adhering to the system's limits.

When compared to PSO and MCS, the implementation of the novel algorithm resulted in a reduction of 6%, 9%, and 2.5%, respectively, in the cost of operation and maintenance for grid-connected, combined, and islanded modes of operation throughout the summer season. In a similar vein, the costs of operation and maintenance dropped by 11%, 2%, and 4%, respectively, throughout the winter season in grid-connected, combined, and islanded modes of operation, respectively. As PSO's functionalities were added, the convergence of hybrid feedback PSO-MCS increased by 8%, resulting in lower microgrid running costs. Convergence occurred in under 10 seconds on average during Python code execution.

In the future, machine learning can be integrated into the system to obtain more precise predictions of solar and wind factors (such as irradiance, cloud cover, wind speed, and so on), and cloud computing can be used to make judgments based on past data of these aspects.

References

[1] W. Yuan, Y. Wang, D. Liu, F. Deng and Z. Chen, "Impacts of Inductor Nonlinear Characteristic in Multiconverter Microgrids: Modeling, Analysis, and Mitigation," IEEE Journal of Emerging and Selected Topics in Power Electronics, vol. 8, no. 4, pp. 3333-3347, Dec. 2020.

[2] Q. Jiang, M. Xue and G. Geng, "Energy Management of Microgrid in Grid-Connected and Stand-Alone Modes," IEEE Transactions on Power Systems, vol. 28, no. 3, pp. 3380-3389, Aug. 2013.

[3] I. U. Nutkani, P. C. Loh, P. Wang and F. Blaabjerg, "Decentralized Economic Dispatch Scheme with Online Power Reserve for Microgrids," IEEE Transactions on Smart Grid, vol. 8, no. 1, pp. 139-148, Jan. 2017.

[4] I. U. Nutkani, P. C. Loh and F. Blaabjerg, "Droop Scheme with Consideration of Oper-ating Costs," IEEE Transactions on Power Electronics, vol. 29, no. 3, pp. 1047-1052, March 2014.

[5] Carvallo, Juan & Schnitzer, Daniel & Lounsbury, Deepa & Deshmukh, Ranjit & Apt, Jay & Kammen, Daniel. (2014). Microgrids for Rural Electrification: A critical review of best practices based on seven case studies.

[6] Aftab Ahmad Khan, Muhammad Naeem, Muhammad Iqbal, Saad Qaisar, Alagan Anpalagan, "A compendium of optimization objectives, constraints, tools and algorithms for energy management in microgrids," Renewable and Sustainable Energy Re-views, Volume 58, 2016, Pages 1664-1683, ISSN 1364-0321.

[7] J. Raya-Armenta, N. Bazmohammadi, J. Avina-Cervantes, D. Saez, J. Vasquez, and J. Guerrero, "Energy management system optimization in islanded microgrids: An over-view and future trends," Renew. Sustain. Energy Rev., vol. 149, pp. 1–20, Dec. 2021.

[8] K. Zhou, S. Yang, Z. Chen, and S. Ding, "Optimal load distribution model of microgrid in the smart grid environment," Renew. Sustain. Energy Rev., vol. 35, pp. 304–310, Jul. 2014.

[9] V. V. S. N. M. Vallem and A. Kumar, "Retracted: Optimal energy dispatch in microgrids with renewable energy sources and demand response," Int. Trans. Electr. Energy Syst., vol. 30, no. 5, pp. 1–27, May 2020.

[10] S. A. Arefifar, Y. A.-R. I. Mohamed, and T. H. M. EL-Fouly, "Optimum microgrid design for enhancing reliability and supply-security," IEEE Trans. Smart Grid, vol. 4, no. 3, pp. 1567–1575, Sep. 2013.

[11] J. Sachs and O. Sawodny, "A two-stage model predictive control strategy for economic diesel-PV-battery island microgrid operation in rural areas," IEEE Trans. Sustain. Energy, vol. 7, no. 3, pp. 903–913, Jul. 2016.

[12] M. Combe, A. Mahmoudi, M. H. Haque, and R. Khezri, "Cost-effective sizing of an AC mini-grid hybrid power system for a remote area in South Australia," IET Gener., Transmiss. Distrib., vol. 13, no. 2, pp. 277–287, Jan. 2019.

[13] S. Ferahtia, A. Djeroui, H. Rezk, A. Houari, S. Zeghlache, and M. Machoum, "Optimal control and implementation of energy management strategy for a DC microgrid," Energy, vol. 238, Jan. 2021, Art. no. 121777.

[14] A. Maulik and D. Das, "Optimal power dispatch considering load and renewable generation uncertainties in an AC–DC hybrid microgrid," IET Gener., Transmiss. Distrib., vol. 13, no. 7, pp. 1164–1176, Apr. 2019.

[15] A. Kumar, M. Alaraj, M. Rizwan and U. Nangia, "Novel AI Based Energy Management System for Smart Grid with RES Integration," IEEE Access, vol. 9, pp. 162530-162542, 2021.

[16] A. Zakariazadeh, S. Jadid, and P. Siano, "Smart microgrid energy and reserve scheduling with demand response using stochastic optimization," Int. J. Electr. Power Energy Syst., vol. 63, pp. 523–533, Dec. 2014.

[17] V. Bui, A. Hussain, and H.-M. Kim, "A multiagent-based hierarchical energy management strategy for multi-microgrids considering adjustable power and demand response," IEEE Trans. Smart Grid, vol. 9, no. 2, pp. 1323–1333, Mar. 2018.

[18] X. Li and S. Wang, "Energy management and operational control methods for grid battery energy storage systems," CSEE Journal of Power and Energy Systems, vol. 7, no. 5, pp. 1026-1040, Sept. 2021.

[19] Y. Li, P. Wang, H. B. Gooi, J. Ye and L. Wu, "Multi-Objective Optimal Dispatch of Microgrid Under Uncertainties via Interval Optimization," IEEE Transactions on Smart Grid, vol. 10, no. 2, pp. 2046-2058, March 2019, doi: 10.1109/TSG.2017.2787790.

[20] Zwe-Lee Gaing, "Particle swarm optimization to solving the economic dispatch considering the generator constraints," IEEE Transactions on Power Systems, vol. 18, no. 3, pp. 1187-1195, Aug. 2003.

[21] N. Augustine, S. Suresh, P. Moghe, and K. Sheikh, "Economic dispatch for a microgrid considering renewable energy cost functions," 2012 IEEE PES Innov. Smart Grid Technol., pp. 1–7, 2012.

[22] M. Bhoye, S. N. Purohit, I. N. Trivedi, M. H. Pandya, P. Jangir and N. Jangir, "Energy management of Renewable Energy Sources in a microgrid using Cuckoo Search Algorithm," 2016 IEEE Students' Conference on Electrical, Electronics and Computer Science (SCEECS), Bhopal, 2016, pp. 1-6.

[23] Premalatha, Kandasamy & Natarajan, A. (2009). "Hybrid PSO and GA for global maximization." Int J Open Prob Compt Math. 2.

[24] J. Zhao, S. Liu, M. Zhou, X. Guo and L. Qi, "Modified cuckoo search algorithm to solve economic power dispatch optimization problems," IEEE/CAA Journal of Automatica Sinica, vol. 5, no. 4, pp. 794-806, July 2018, doi: 10.1109/JAS.2018.7511138.

[25] "Time of Use," opuc.on.ca. http://www. https://www.opuc.on.ca/time-of-use/ (accessed: March. 15, 2022).

[26] Zwe-Lee Gaing, "Particle swarm optimization to solving the economic dispatch considering the generator constraints," IEEE Transactions on Power Systems, vol. 18, no. 3, pp. 1187-1195, Aug. 2003.

[27] X. -S. Yang and Suash Deb, "Cuckoo Search via Lévy flights," 2009 World Congress on Nature & Biologically Inspired Computing (NaBIC), 2009, pp. 210-214.

[28] M. Y. Ali, F. Khan and V. K. Sood, "Energy Management System of a Microgrid using Particle Swarm Optimization and Wireless Communication System," 2018 IEEE Electrical Power and Energy Conference (EPEC), 2018, pp. 1-7.

6

Topological Investigations of Grid Integrated Renewable Energy Systems for Power Quality Improvement

Shashank Shrivastava[1], Shitanshu Kumar[1], Shailendra Kumar[2], Sanjeev Singh[1], Suresh Kumar Gawre[1], D. Giribabu[1]

[1]Department of Electrical Engineering, MANIT Bhopal, India
[2]Department of Electrical Engineering IIT Bhilai, India
Email: 741shashank@gmail.com, shitanshushahi@gmail.com, skumaree@iitbhilai.ac.in, sschauhan.sdl@gmail.com, sgawre28@gmail.com, dgiribabu208@gmail.com

Abstract

Photovoltaic (PV) systems have grown in importance over the last decade. Due to their unique designs and growing commercial interest, grid-connected single-phase transformerless (TL) solar inverters are now being explored further. There are several benefits over transformer galvanic isolation-based inverters, including cheaper prices, lower weight, smaller volume, higher efficiency, and less complexity. In this chapter, transformerless "grid-connected" inverters with negative, positive, and zero cycle operations are explored. A detailed examination of various topologies is carried out. Simulated results are shown to verify the effectiveness of the systems.

Keywords: Renewable energy resources (RESs), leakage current, solar photovoltaic, transformerless, multilevel, power converter

6.1 Introduction

Traditional power-generating fossil-based resources, such as coal and petroleum, are unsustainable. In the previous decade, grid connectivity has improved with the widespread use of renewable energy resources such as hydro, wind, photovoltaic, geothermal, biomass, tidal power, and thermal power generation [1]. Solar power generation is becoming increasingly important in many nations, and as a result, the number of solar PV installations is rapidly increasing [2]. For instance, in the stated policies scenario (STEPS) – or even sooner in the sustainable development scenario – solar power is predicted to grow at a fast rate in India, eventually displacing coal as the country's primary source of energy. Out of which coal accounts for over 70% and in other possibilities, the transition must be substantially faster. This huge transition is occurring due to legislative desire, and specifically the objective is to increase the renewable contribution by 2030 [3–4].

Some interface devices are required for the grid connection of these sources like PV arrays [5]. Progress in semiconductor technology has made it possible to convert clean, pollution-free, and limitless solar PV energy into reliable, efficient, and sustainable power [6–8]. Many grid-connected PV power converters are now in use because of this trend. An inverter is used to convert DC power from a PV module to AC power. A DC-DC converter and either a voltage source inverter (VSI) or current source inverter (CSI) are commonly used to connect PV systems to the grid. A sophisticated control system is employed to get the best performance and desired results from the system. During the control process an inverter is used to control the DC link voltage so that the flow of both active and reactive powers to the grid can be controlled easily [8].

The advances in the power electronics industry have led to the H-bridge inverter design which can be implemented in transformerless (TL) fashion [9–12]. This topology results in more efficient, better performance, and low electromagnetic interference (EMI)/common-mode (CM) emissions. Due to the lack of electrical separation between PV panels and the grid in a transformerless inverter topology (TLIT), leakage currents and high-frequency common-mode voltages might occur. It flows via the PV panel's parasitic capacitance. This should be avoided at all costs. Therefore, many TLIT-based solar PV inverters have been proposed in the literature to minimize earth leakage currents and improve system efficiency [13–15]. These papers also compare the highly efficient and reliable inverter concept (HERIC)-based design to alternative transformerless inverter topologies that include a DC-DC converter and only function when the PV array voltage falls below a certain value.

The upgraded topology provides the following features:

a) It improves the overall efficiency of the system by minimizing the two-step conversion process.

b) The multi-level inverter output serves to minimize grid voltage and output current distortion.
c) Transformerless (TL) PV systems are suitable for grid connections because the level of CM leakage current is minimized [16, 17].

In a comparative study, the efficiency of three TLIT topologies has been evaluated (HERIC-based, HERIC-enhanced topology, and H5-enhanced topology). When the PV generator output voltage changes or dips below the peak voltage, the DC-DC converter is activated, forcing power to move from the PV array to the second DC-link capacitor, resulting in the complete DC-link voltage exceeding the peak network voltage. As a result, multi-level operation of the circuit is demonstrated, including five-level output voltage while running at low PV voltage. Finally, the main purpose of these studies is to select a suitable topology that meets with the requirements.

6.2 Types of Topologies and Their Modulation Approaches

The converter topologies related to grid-integrated renewable energy sources along with their modulation techniques are discussed next.

6.2.1 H-Bridge-based inverter structures

Earlier, this arrangement was made by using thyristors, commonly known as force-commutated semiconductor devices. Either two or four controlled devices are required to build this topology; they may be used for both conversion and inversion modes, making it a flexible circuit. Figure 6.1 depicts the fundamental topology of a full bridge circuit.

Single-stage inverters do not require a maximum power point tracking (MPPT) DC-DC converter. The half-bridge (HB) grounds the midpoint of the capacitive divider to prevent leakage currents (LCs) and ensure common mode (CM) voltage management. The parasitic capacitance of the PV module provides a conduit for leakage current to flow through the system [18]. The defining aspect of this topology is its low cost. It has two levels of output voltage waveforms. Additionally, the current is distorted, resulting in significant distortion, which is also a drawback of the topology [19]. Calais et al. [20] have developed an inverter that considers factors such as PV array incentive, which includes ground resistance, stress, power rating of the system, and module count for grid-connected solar power PV systems. The H-bridge is arranged into 5-levels in grid-tied PV systems.

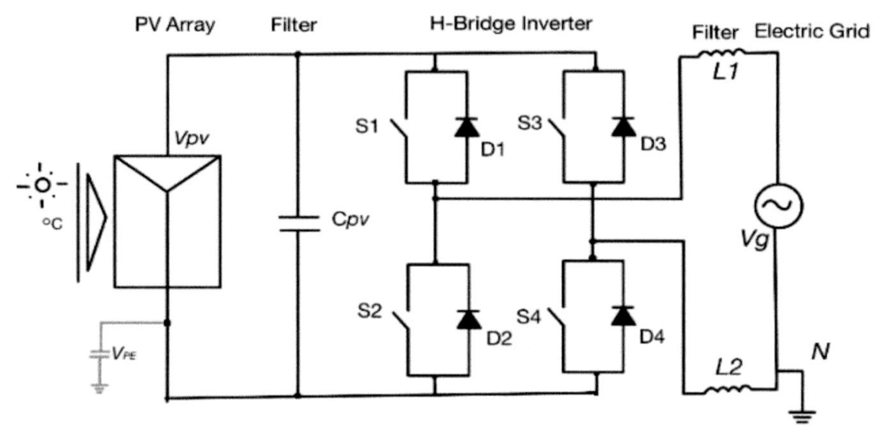

Figure 6.1 General circuit of PV inverter [13].

6.2.2 Modulation strategies

Before continuing further, we provide some information about modulation strategies which are generally preferred, including their pros and cons. The analysis has also been made, which modulation techniques can be used for transformerless inverter topologies. Further, these techniques work in combination with different inverter topologies such as the combination of hybrid structure with H5 inverter topology.

The modulation techniques are categorized as:

- Bipolar structure
- Unipolar structure
- Hybrid structure

6.2.2.1 Bipolar modulation

Two-level modulation is another name for the bipolar modulation technique. The diagonal switches are activated as S_1 with S_4 or S_2 with S_3. The converter is responsible for obtaining the AC output voltage, as shown in Figure 6.2 (a,b). The following functions are included in this converter:

- The diagonal switches S_1 with S_4 or S_2 with S_3 are synchronously switched on at a high frequency (more than 10 kHz).
- The state for zero voltage is not possible at the output end.

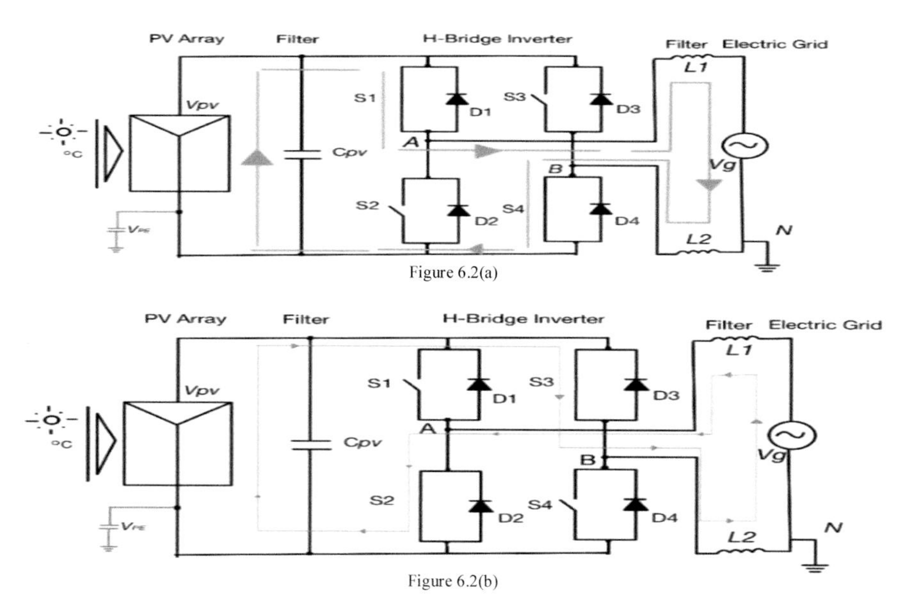

Figure 6.2 PV inverter bipolar modulation operation for (a) positive and (b) negative cycle [20].

6.2.2.1.1 Advantages

- The electromagnetic interference (EMI) and leakage current (LC) are very low because voltage to ground V_{PE} has only grid frequency component.

6.2.2.1.2 Disadvantages

(a) Core losses in the output filter are high, as there is an interchange of reactive power between L_{12} and C_{PV}, resulting in an overall low efficiency of around 95%.

(b) The filtering circuit needed is larger because the switching ripple is equivalent to the $1\times$ switching frequency.

(c) Bipolar voltage fluctuations produce higher core losses.

6.2.2.1.3 Analysis

It has a benefit of low leakage current, but it has reduced efficiency and, due to this reason, it isn't suitable for transformerless inverter PV systems.

6.2.2.2 Unipolar modulation

Three-level modulation is another name for this approach. Each leg's switching signal is determined by its respective reference signal. As shown in Figure 6.3, the AC voltage may be generated for both positive and negative output currents. The features of this converter are:

- Leg A and Leg B get high-frequency switching with reflected sinusoidal reference.
- To produce a zero-state output voltage, use the switches in the following order: S_1, S_3, or S_2, S_4.

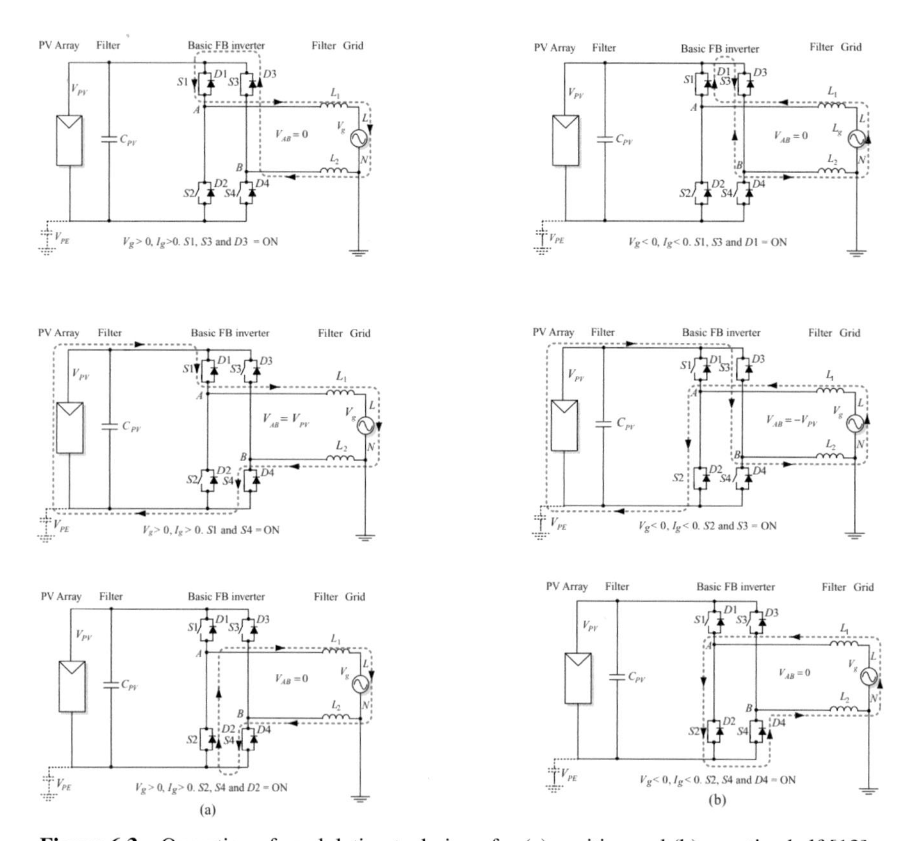

Figure 6.3 Operation of modulation technique for (a) positive and (b) negative half [13].

6.2.2.2.1 Advantages

- The efficiency is good, up to 98%, due to lower losses during zero voltage situations.

6.2.2.2.2 Disadvantages

- The leakage currents (LC) and EMI are very high.

6.2.2.2.3 Analysis

Despite its improved efficiency and reduced filter circuit requirements, this modulation approach is not suited for transformerless PV systems due to the high-frequency content of voltage to ground (V_{PE}).

6.2.2.3 Hybrid modulation

Using the hybrid modulation approach [21, 22], one leg is activated at a higher frequency, while the other leg is activated at grid frequency. Positive and negative output currents may be generated by manipulating the AC voltage, as shown in Figure 6.4. The following characteristics are at the core of this converter.

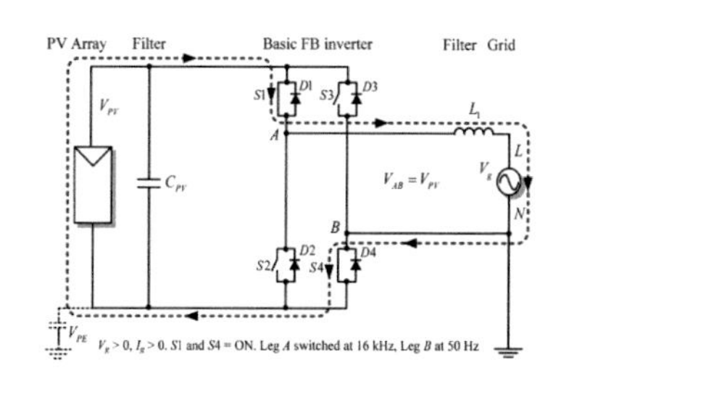

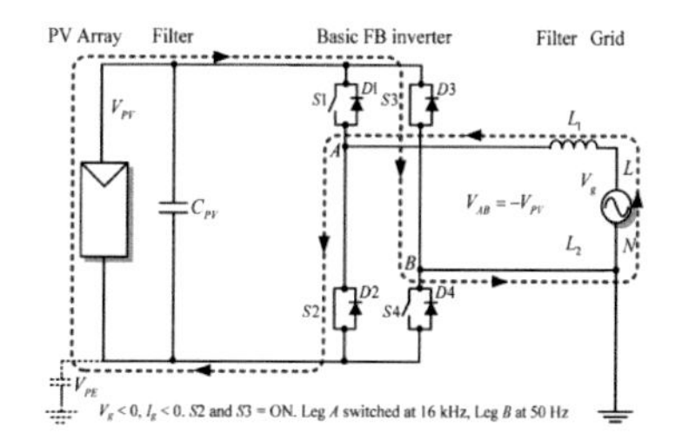

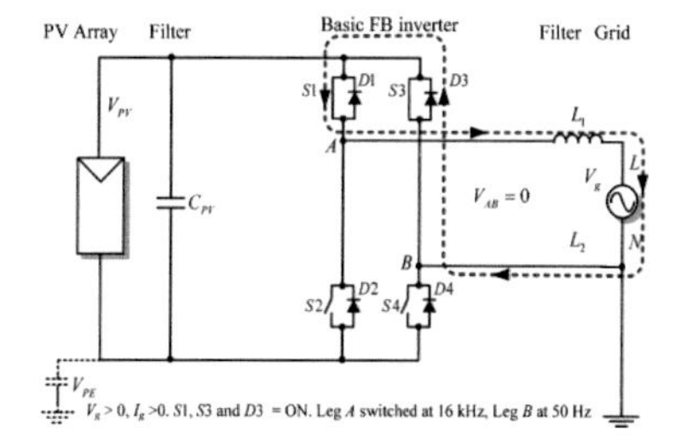

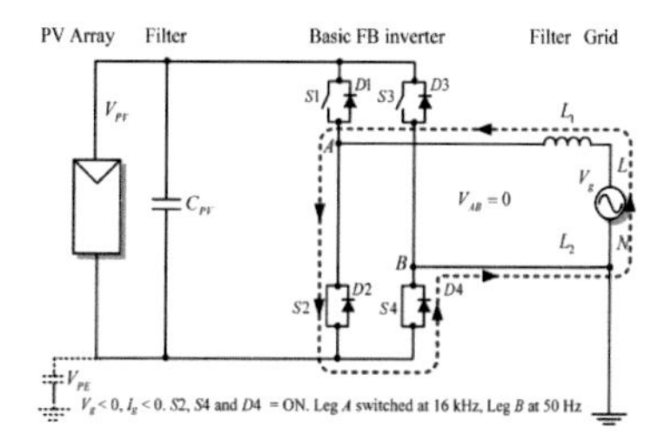

Figure 6.4 Hybrid switching scheme (HM) for (a) (+)ve half and (b) (−)ve half.

- Leg A receives a higher PWM frequency, while Leg B receives a fundamental switching.
- Zero switching states are achieved when S_1, S_2, or S_3, S_4 are turned on.

6.2.3 H5 structure

This architecture is known as the H5 and was patented by SMA Solar India Pvt. Ltd. in 2005. This is a modified H-bridge with a DC link and an additional 5th bus, as shown in Figure 6.5 [23]. The additional switch has two purposes. Reactive power is not exchanged between C_{pv} and $L_{1(2)}$ when the voltage is zero, improving efficiency and avoiding the high-frequency components of V_{PE} by separating the PV module from the grid. Figure 6.6 shows the generation of alternating current in negative and positive switching states. This design underscores the advantages of an enhanced SiC-based H5 topology over nonoptimized Si-based competitors for power generation. The underlying properties of this converter are [24, 25]:

- When switches S_4 and S_2 are on and S_5 is OFF, voltage states with two zero outputs are available.
- High-frequency switching is used for S_2, S_4, and S_5, whereas grid frequency is used for S_3 and S_1.

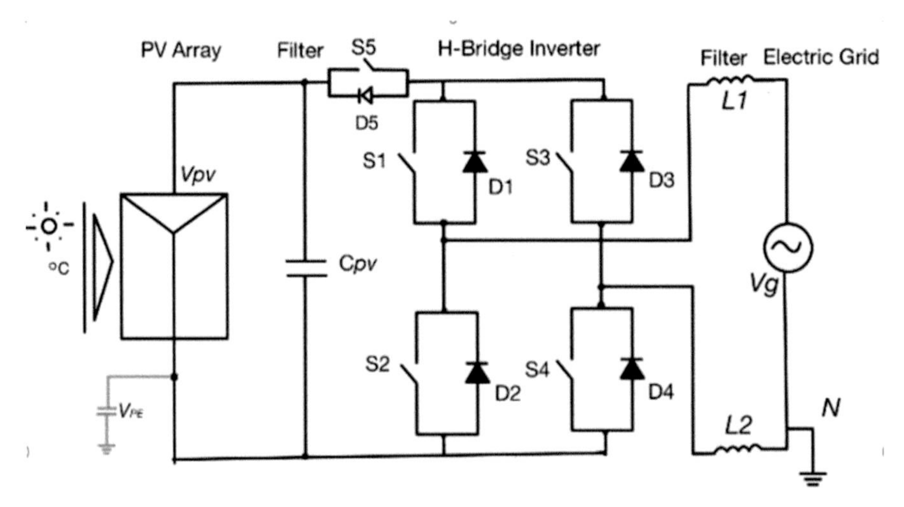

Figure 6.5 H5 (SMA) inverter topology.

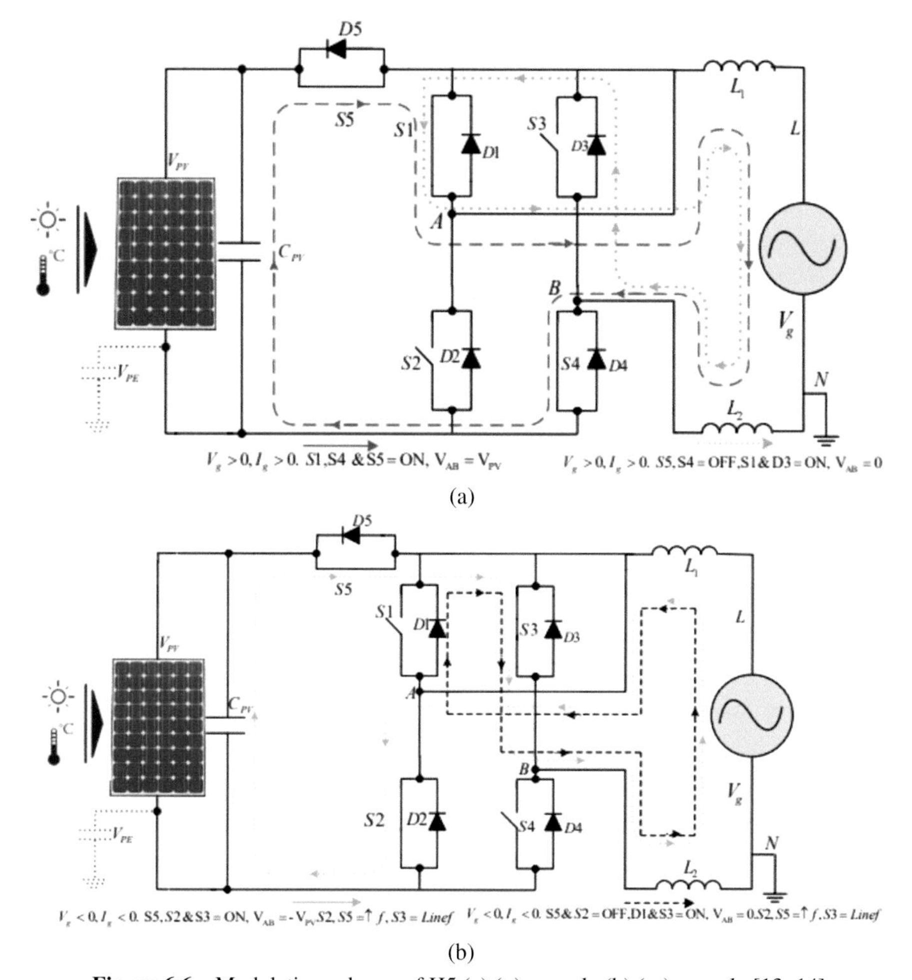

Figure 6.6 Modulation scheme of H5 (a) (+)ve cycle (b) (−)ve cycle [13, 14].

6.2.3.1 Advantages

(a) Due to its unipolar voltage fluctuation, the H5 inverter's core losses are reduced.
(b) The inverter's efficiency rises to 98%.
(c) The voltage to ground ($V_{PE's}$) single-grid frequency component reduces EMI filtering, thereby reducing peaks of leakage current [26].

6.2.3.2 Disadvantages

As three switches are conducting, conduction losses are larger during active operation. But the overall high efficiency is unaffected. There is a need for one more switch.

6.2.3.3 Analysis

The H5 topology combines the benefits of full-bridge and hybrid modulation schemes. The high-frequency component of the V_{PE} is avoided by adding a switch to disconnect the grid from the PV module when the voltage is zero. High efficiency (typically 98%), low EMI, and minimal output filtering requirements make this technique ideal to PV systems. It is part of SMA's Solar India Pvt. Ltd. products (Sunny Boy) [26].

6.2.4 HERIC structure

In 2006, Sunways-tech patented this design, shown in Figure 6.7. Because of its reliability and efficiency, it is one of the most used inverter systems. On the AC side, there is a bypass leg added to two back-to-back insulated gate bipolar transistors (IGBTs). By adding a fifth switch, the AC bypass accomplishes the same thing as the H5 design. $L_{1(2)}$ and C_{pv} do not exchange reactive power when voltage is zero. To eliminate the higher frequency

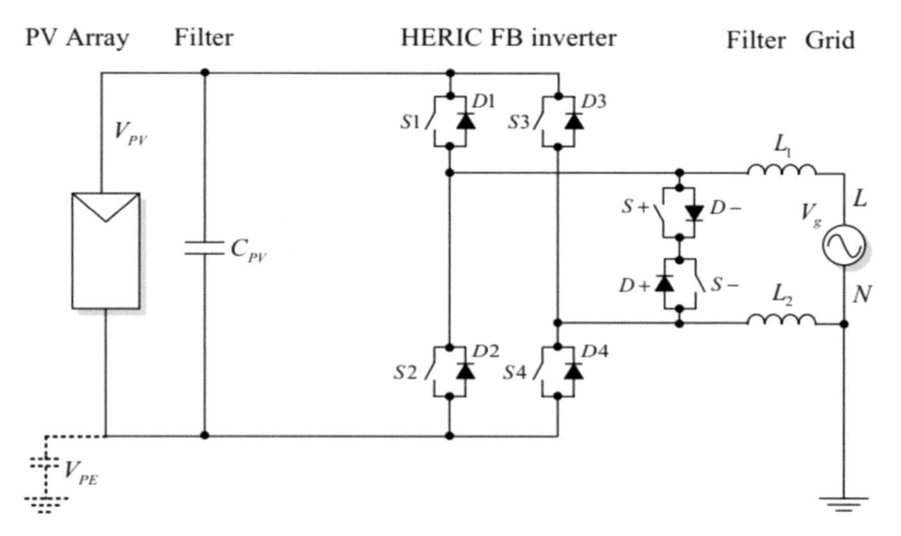

Figure 6.7 HERIC inverter topology [34].

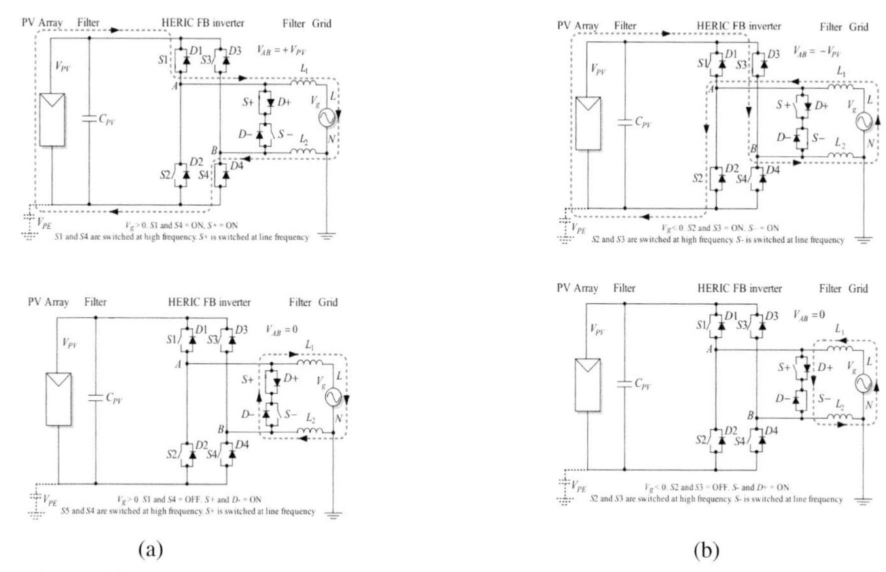

Figure 6.8 Switching states of the Sunways HERIC (a) (+) ve cycle (b) (−) ve cycle [34].

component of V_{PE}, the grid must be disconnected from PV panels while the voltage is zero [27–32].

Positive and negative switching states are shown as generating alternating current (AC) in Figure 6.8. Assuming that the bridge is not off, there are two conceivable zero output voltage states:

(a) when S+ and S− both come on (in the case of the bridge being off) or;

(b) when S1− or S2− or S3 are switched on at high frequency while S+ and (S−) is switched on at grid frequency.

6.2.4.1 Advantages

The following are some of the advantages of the highly efficient and reliable inverter concept HERIC topology:

(a) The unipolar voltage variation reduces the core loss component.

(b) At zero voltage, reactive power interaction is not there between C_{PV} and $L_{1(2)}$, yielding a 97% efficiency. Furthermore, because V_{PE} only includes line frequency components and no switching frequency components, EMI filtering needs are minimal, and leakage current peaks are minimal.

6.2.4.2 Disadvantages

There is a requirement for two additional devices.

6.2.4.3 Analysis

Adding a zero-voltage level to the HERIC architecture improves its efficiency. For practical uses, it has minimal EMI and low filtering needs due to its great efficiency. Sunways sells the AT series (2.7–5.0 kW) of solar panels, which has a maximum efficiency of 95.6%. The operational behaviour of HERIC and H5 is essentially comparable since the grid should be disconnected from the PV generator on both the DC and AC sides during zero-voltage switching.

6.2.5 Improved HERIC structure

This inverter architecture is shown in Figure 6.9. It combines a HERIC inverter design (Figure 6.10) with six switches (S_1–S_6) with two more switches (S_7 and S_8). This circuit is meant to operate in two different modes, one in three levels and the other in five levels.

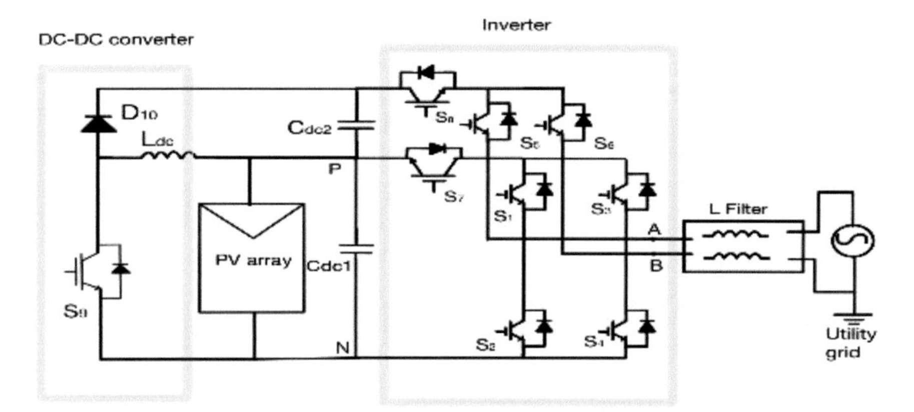

Figure 6.9 Improved HERIC structure (diodes of S1–S9 are denoted as D1–D9).

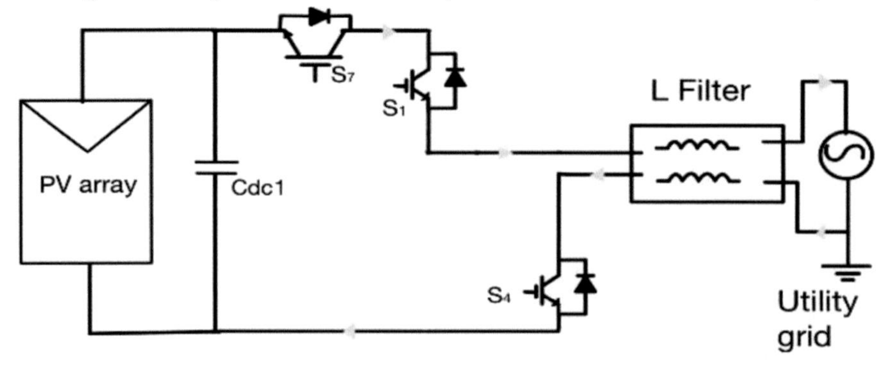

Figure 6.9 (a). Operation in mode 1/mode 3.

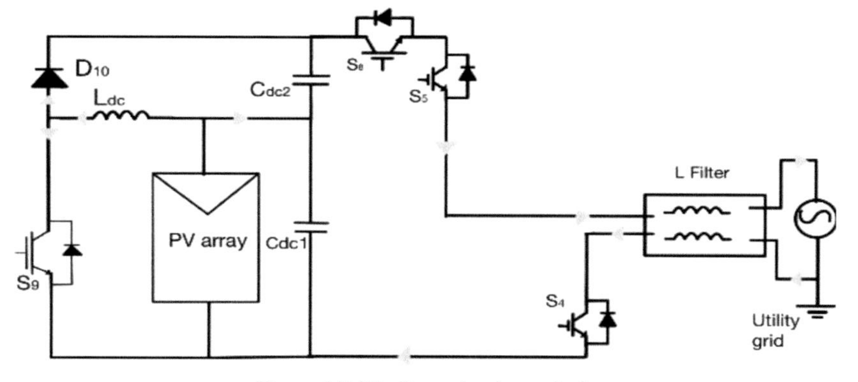

Figure 6.9 (b). Operation in mode 2.

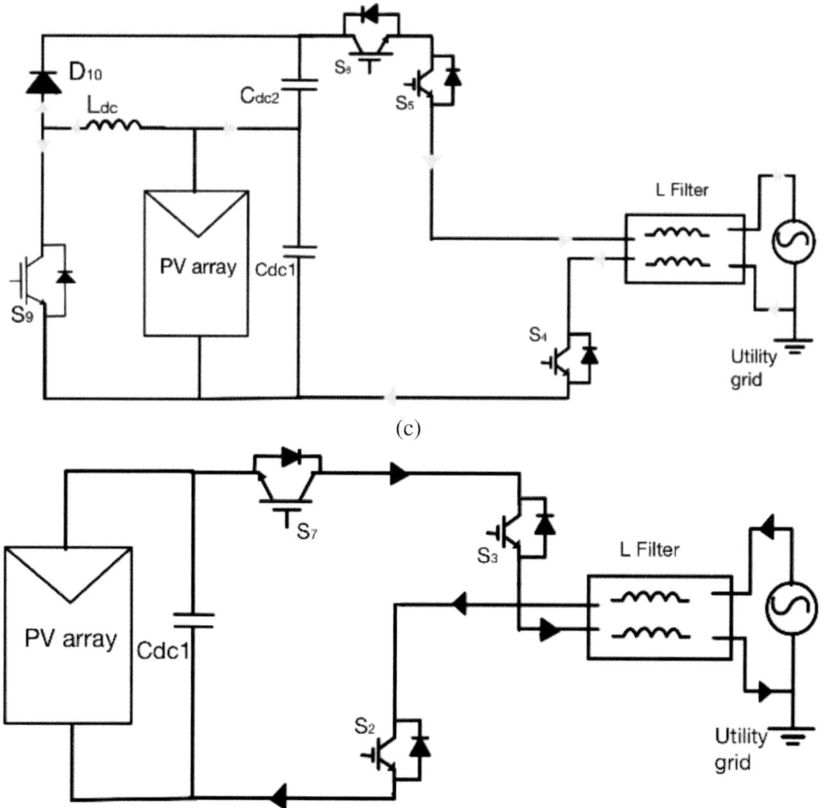

(c)

Figure 6.9 (c). Operation in all modes (Figure 6.9 (d). Operation in mode 4/mode 6 direction depends on the mode of operation).

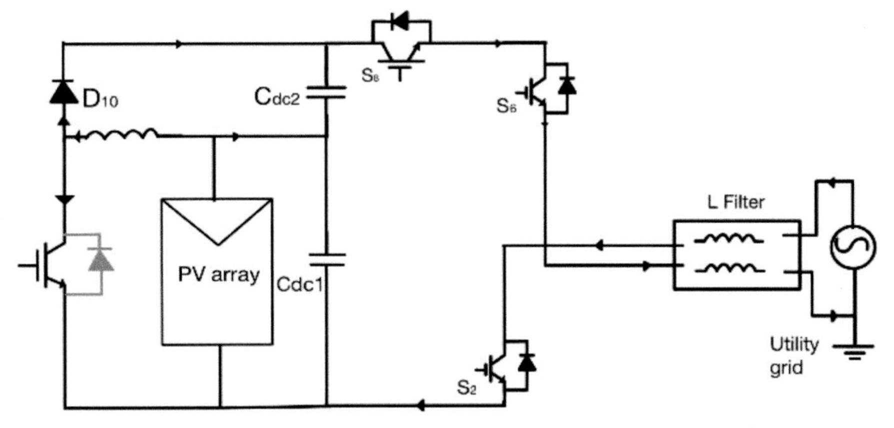

Figure 6.9 (e) Active state in mode 5.

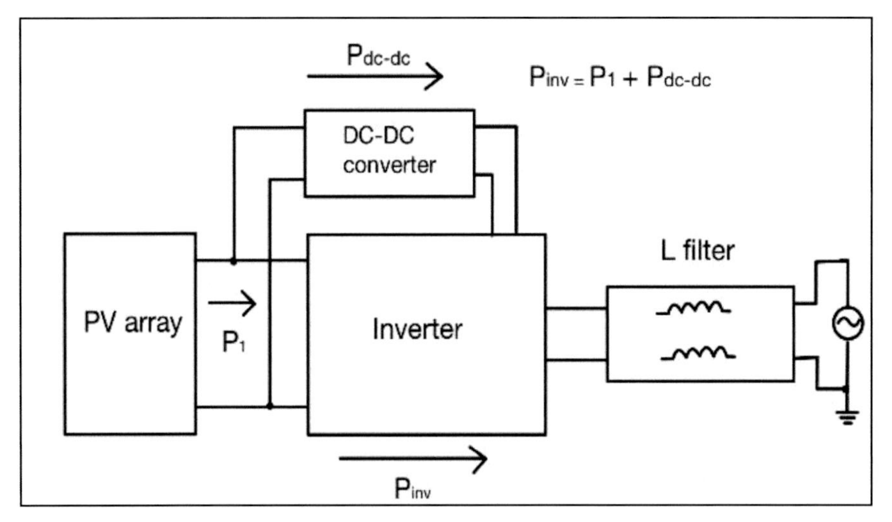

Figure 6.10 Improved HERIC converter structure.

6.2.5.1 Working of different modes

- Activation of switch S_7 and deactivation of switch S_8 occurs when DC voltage surpasses maximum utility voltage. Thus, a HERIC inverter may be created using this circuit.
- During this mode of operation, the DC-DC converter is turned off. This enables the 3-level operation mode as shown in Figure 6.11.

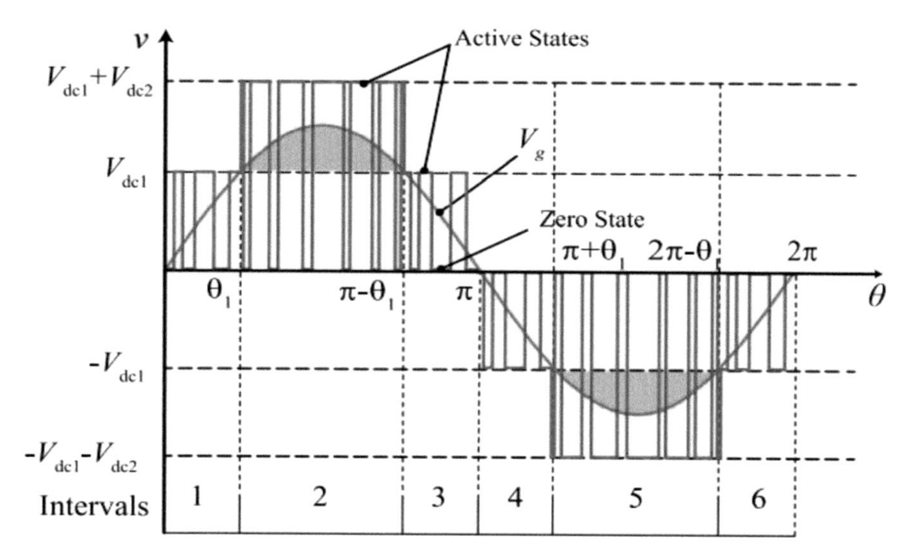

Figure 6.11 Switching states of HERIC structure [25].

- When the PV array output voltage falls below the peak grid voltage, the DC-DC converter is activated. Because of this, the second DC-link capacitor C_{dc2} receives power from the PV array and raises the voltage of the whole DC-link system over the maximum grid voltage (V_{dc1} + V_{dc2}).
- When the PV array voltage is greater than the instantaneous grid voltage, the voltage across C_{dc1} is "V_{dc2}" and is solely utilized by the device.
- Current is drawn from the PV, when the PV voltage V_{dc1} is less than the immediate grid voltage. As a result, the DC-DC converter only passes a tiny part of the PV energy.
- The five-level technique functioning is explained below. Figure 6.9 shows the reference utility voltage is bifurcated on six distinct operating periods.

Period 1 ($0 < \theta < \theta_1$):

- An active (V_{dc1}) and a zero state are used to achieve output voltage.
- To activate the switch, one needs to turn on the switches S_1, S_4, and S_7. Figure 6.9 (a–c) depicts the route of current flow in this condition (a).

The boost converter is not turned on. Switches S_5 and S_7 are kept on constantly throughout this time, preventing switching losses in these devices [33–36].

Period 2 ("$\theta_1 < \theta < \pi - \theta_1$"):

- Throughout this period, the inverter output voltage is shifting between zero and active state (V_{dc1} + V_{dc2}).
- Switching on switches S_5 and S_6, as illustrated in Figure 6.9, achieves zero state (c).
- As illustrated in Figure 6.9, switches S_4, S_5, and S_8 are switched on to achieve the active state of (V_{dc1} + V_{dc2}). The boost converter regulates the MPPT as it uses S9, D10, and LDC components (S_9, D_{10}, L_{dc}).

Period 3 ("$\pi - \theta_1 < \theta < \pi$"):

The functioning in this period is like the period 1. The two states that are used are the switching on (presented in Figure 6.9 (a)) and zero state (presented in Figure 6.9 (c)).

Period 4 ('$\pi < \theta < \pi + \theta_1$'):

Throughout this period, the reference signal fluctuates between 0 and V_{dc1}. As seen in Figure 6.9, the zero state has been achieved. V_{dc1}, S_2, S_3, and S_7 are turned on in the active state, as indicated in Figure 6.9 (d). This setting disables the DC-DC converter.

Period 5 ('$\pi + \theta_1 < \theta < 2\pi - \theta_1$'):

Both the modes are active in this region.

Period 6 ('$2\pi - \theta_1 < \theta < 2\pi$'): The topology's functioning in this time is identical to that in interval 4. There are two states: active and zero.

6.2.6 Parameter comparison

Comparison is made on the basis of various aspects such as number of active and passive elements used on transformerless inverter topology. Effectiveness of different topologies can be easily studied via Table 6.1 comparison.

Table 6.2 includes comparison of the pros and cons of various topologies

Table 6.1 Comparative assessment of SPGC TLIT

Inverter types/performance indices	H5 topology	HERIC topology	HERIC-based modified
Input capacitor	*1*	*1*	*1*
Switches	*5*	*6*	*9*
Diodes	*0*	*2*	*1*
O/p voltage level	*3*	*3*	*3 and 5*
Number of trackers	*1*	*1*	*1*
Leakage current	*Very low*	*Very low*	*Very low*
Max efficiency (%)	*98.5*	*98.6*	*98.4*
Cost	*High*	*Moderate*	*High*

Table 6.2 Comparison of different TLIT.

Reference	Topologies	Leakage current	Switches	Disadvantages	Advantages
[30]	Half bridge	Moderate	2T	High voltage stress in DC-link	Low cost
[31]	Full bridge	Moderate	4T	High leakage current i	-
[32]	H5	Low	5T	Unbalanced switching	Low component
[33]	HERIC	Low	6T + 2D	Extra devices required	Line frequency leakage current
[?]	HERIC modified	Low	9T + 1D	Extra devices required	Higher efficiency and low CM leakage current

6.2.7 Comparison of losses in these topologies

- The improved HERIC-based architecture has a greater efficiency at lower voltage levels, as seen in this graph (Figure 6.12 and Figure 6.13).
- The DC-DC converter processes a tiny fraction of PV power. Even when the PV module voltage is low, most of the electricity is transmitted straight to the inverter, which increases efficiency.
- The conduction losses in the modified topology are higher than in the HERIC topology since extra switches/diodes are added to the conduction channel in this case.
- At a PV voltage of 280 V, the suggested design has been proven to have lower switching and hence total losses than the HERIC + boost topology, which uses partial power processing rather than complete power processing with a multilevel topology.
- It has been observed that when the PV voltage is higher than the AC voltage peak, both H5 and HERIC have higher efficiency.
- The recommended or modified topology-based inverter, on the other hand, surpasses the other two topologies in terms of efficiency when the PV voltage decreases or at lower voltage levels.
- This is because the recommended topology's DC-DC converter only needs to handle a portion of the entire power, unlike the H5 or HERIC + boost topologies.

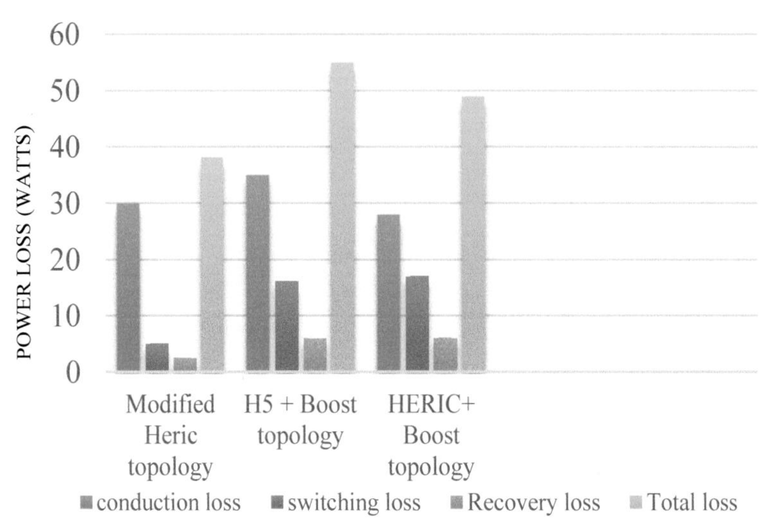

Figure 6.12 For three distinct topologies, the total power loss is: for a PV voltage of 280 V and a power of 2.5 kW [36, 37].

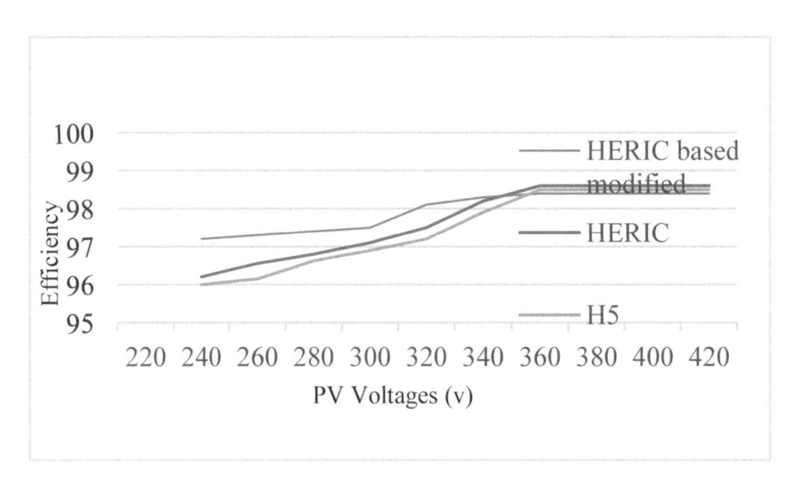

Figure 6.13 Modified HERIC-based topology with H5 and HERIC + boost topologies at various voltage levels are compared in terms of efficiency (simulated) at 2.5 kW of power output.

6.2.8 System configuration under study

Figure 6.14 depicts a three-phase two-stage PV system. For MPPT, a boost DC-DC converter is utilized using incremental conductance technique. The

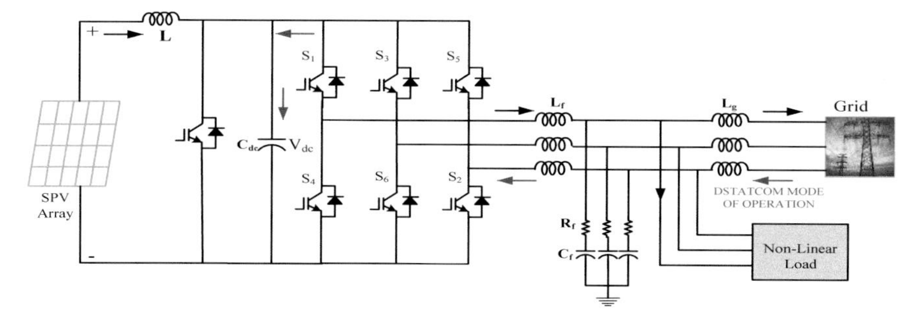

Figure 6.14 System topology.

VSC is controlled using adaptive dq-based control technique. It improves the dynamic and steady state operation of the SECS. When PV power is available the power is supplied to the grid and load. In DSTATCOM mode of operation grid supply power to the load. At the input terminal of VSC, the DC link voltage is connected which is regulated using a cascaded PI controller. At PCC, the VSC is linked with grid by an inductive filter to eliminate the current harmonics. An RC filter is used to reduce the ripples of voltages which are caused by VSC switching.

6.2.9 Control solutions for renewable energy system

Here, the control related to grid interactive renewable energy system is discussed. The control is based on dq0 to abc transformation and vice-versa. The mathematical evaluation is given in this section. The main goal is to improve the grid current power quality by functioning as DSTATCOM and supporting the common DC bus voltage by absorbing the active power.

The control algorithm is depicted in Figure 6.15. As feedback signals, the load currents (i_{La}, i_{Lb}, i_{Lc}), PCC voltages (V_{ga}, V_{gb}, V_{gc}), and DC bus voltage (V_{dc}) are monitored.

The abc frame's load currents are first transformed to the dq0 frame of reference as similar to the *p–q* theory. Current components in α–β coordinates are generated, and using θ as a transformation angle, these currents are transformed from α–β to *d–q* frame (Park's transformation).

It is defined as:

$$x_{dqo} = Kx_{abc} = \sqrt{\frac{2}{3}} \begin{bmatrix} \cos\theta & \cos\left(\theta - \frac{2\pi}{3}\right) & \cos\left(\theta + \frac{2\pi}{3}\right) \\ -\sin\theta & -\sin\left(\theta - \frac{2\pi}{3}\right) & -\sin\left(\theta + \frac{2\pi}{3}\right) \\ \frac{\sqrt{2}}{2} & \frac{\sqrt{2}}{2} & \frac{\sqrt{2}}{2} \end{bmatrix} \begin{bmatrix} x_a \\ x_b \\ x_c \end{bmatrix}$$

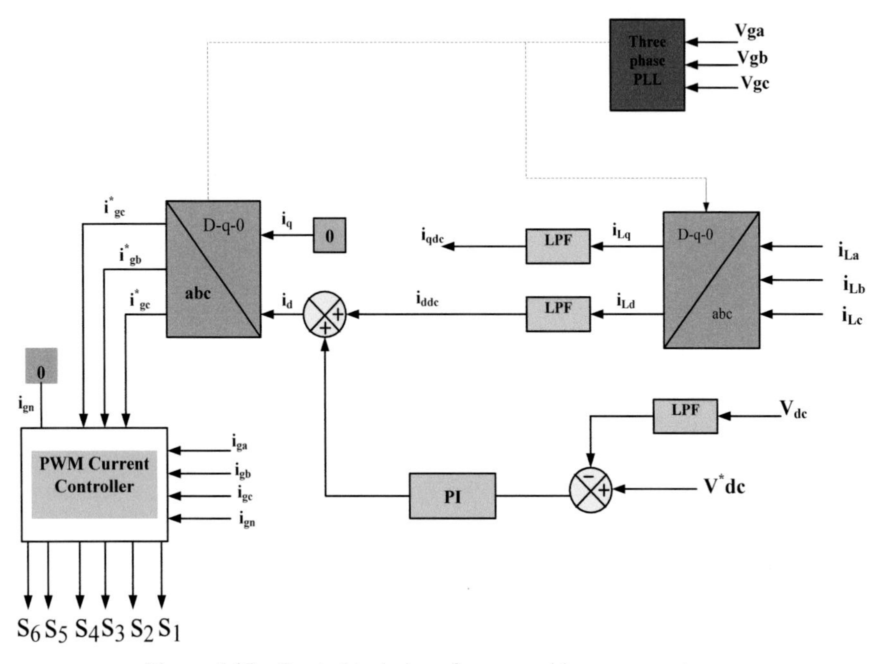

Figure 6.15 Control technique for renewable energy systems.

Inverse Park transformation

$$x_{abc} = K^{-1}x_{dqo} = \sqrt{\frac{2}{3}}\begin{bmatrix} \cos\theta & -\sin\theta & \frac{\sqrt{2}}{2} \\ \cos\left(\theta - \frac{2\pi}{3}\right) & -\sin\left(\theta - \frac{2\pi}{3}\right) & \frac{\sqrt{2}}{2} \\ \cos\left(\theta - \frac{2\pi}{3}\right) & -\sin\left(\theta + \frac{2\pi}{3}\right) & \frac{\sqrt{2}}{2} \end{bmatrix}\begin{bmatrix} x_d \\ x_q \\ x_o \end{bmatrix}$$

Over PCC voltages, angles are determined using a three-phase PLL. From PCC voltages, a PLL signal is generated, which is used to generate fundamental unit vectors for converting detected currents to the dq0 reference frame. LPFs in the SRF controller extract DC quantities, separating non-DC quantities (ripple) from the reference signal. DC (fundamental) and ripple (negative sequence and harmonics) components make up the *d*- and *q*-axis currents.

$$\begin{aligned} i_{Ld} &= i_{\text{ddc}} + i_{\text{dac}}, \\ i_{Lq} &= i_{\text{qdc}} + i_{\text{qac.}}. \end{aligned} \tag{6.1}$$

The AC mains must deliver the mean value of the direct-axis component of the load current, as well as the active power component of the current, to maintain the DC bus voltage and meet the losses (i_{loss}), according to this control approach. The current (i_{loss}) for addressing the losses is taken from the PI controller's output at the DC bus voltage:

$$i_{\text{loss}}(n) = i_{\text{loss}}(n-1) + K_{p1}\{V_{dd}(n) - V_{dd}(n-1)\} + K_{id}V_{dc}. \quad (6.2)$$

K_{pd} and K_{id} are the proportional and integral gain constants respectively, of the DC bus voltage PI controller, and DC voltage at the *n*th sampling moment. As a result, the reference supply current's amplitude is

$$i^* = iddc + i\,\text{loss}\,n \quad (6.3)$$

The reference supply current must be in phase with the PCC voltage, but without any zero-sequence components. It is therefore obtained by reversing Park's transformation with i^*q and i^*0 set to zero. The reverse Park's transformation is used to convert the dq0 currents back into the reference supply currents to generate the switching pulses.

6.3 Results and Discussion

In this section, simulated performances of the renewable energy system under different operating condition with nonlinear loads are discussed. The MATLAB-simulated results have been discussed under various conditions such as insolation variations, load unbalancing, grid voltage sag and swell conditions. Here, the internal signal response is also taken into consideration to describe the response of the system.

6.3.1 Response of the internal signals

Figure 6.16 exhibits the results of the internal signals of system in load changing condition. At time instant 0.3 s, the load is unbalanced by making phase "a" current zero and again at time instant 0.5 s it is changed to the normal operation. During this period the fundamental extracted current also changes which indicates the response of the system is good. However, the reference current is unaffected which shows the steady state stability of the system using dq-based control.

6.3.2 Response of system under insolation variations

Figure 6.17 exhibits the results of the system in changing insolation. At time instant 0.2 s, the insolation is changed from 1000 Wm^{-2} to 500 Wm^{-2} and again at time instant 0.4 s it is changed back from 500 Wm^{-2} to 1000 Wm^{-2}.

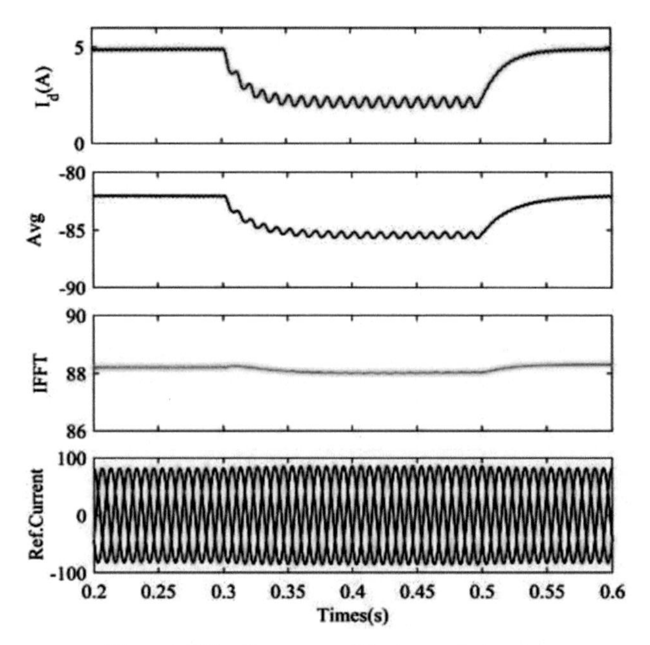

Figure 6.16 Response of the internal signals.

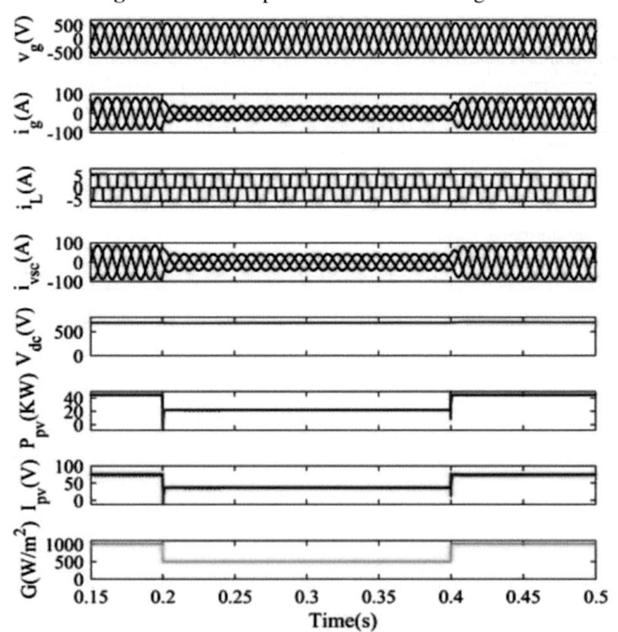

Figure 6.17 Simulated result at insolation variations.

During this period the load current is constant. However, PV power (P_{pv}) and PV current (I_{pv}) increases because of this, grid current also increases. Thereby, injecting more power to the grid, as the grid voltage and current are out of phase with unity power factor.

6.3.3 Operation of system in DSTATCOM mode of operation

Figure 6.18 shows the results when insolation is present and absent (DSTATCOM). In Figure 6.19, at time instant of 0.25 s when insolation is reduced to zero, the load current remains constant and grid current is reduced. At time instant of 0.5 s when insolation is increased from zero to 1000 Wm^{-2} the grid current also gets increased while load current remains constant.

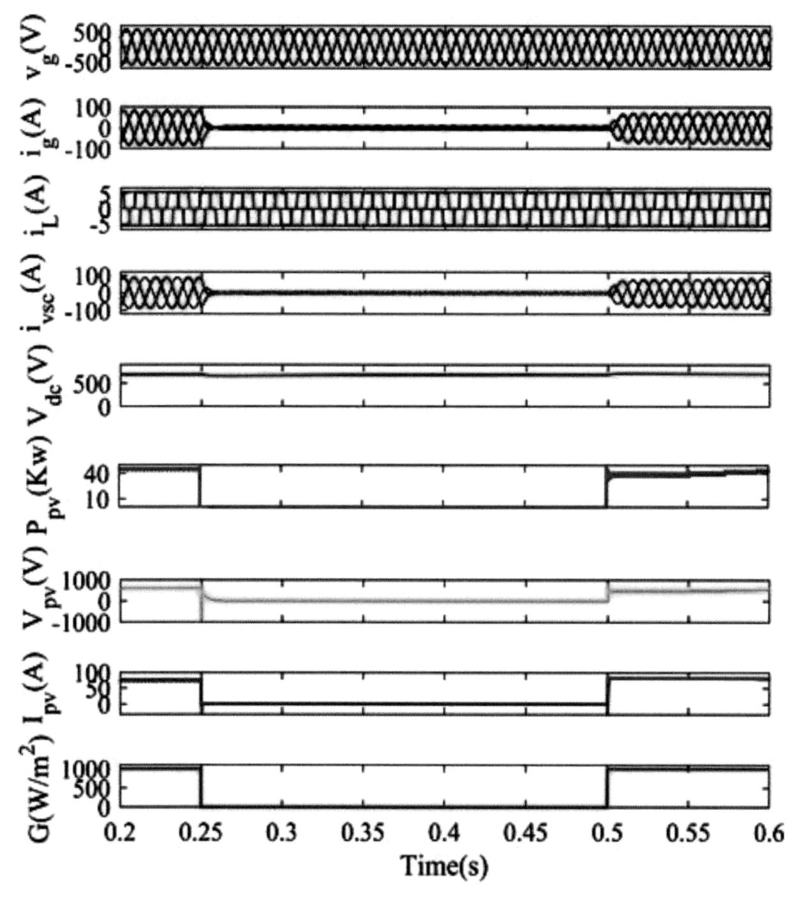

Figure 6.18 Response under DSTATCOM mode of operation.

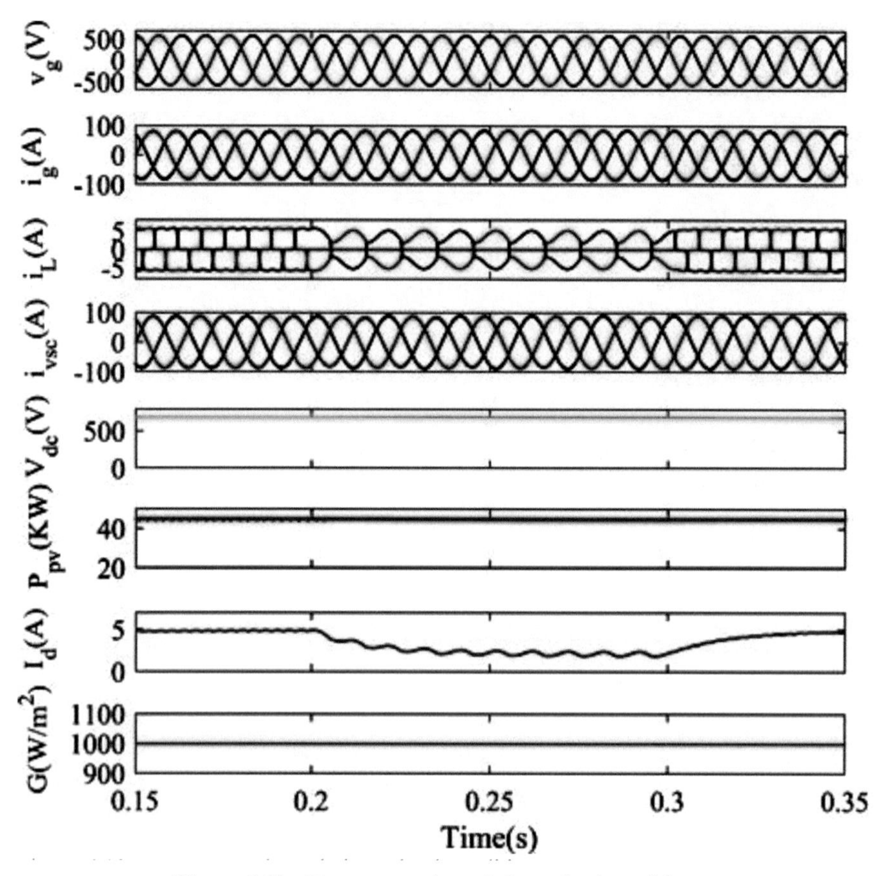

Figure 6.19 Response under unbalance load conditions.

6.3.4 Operation of system under load disturbance

Figure 6.20 exhibits the results when load unbalancing occurs in the system. When the load current is unbalanced the grid current is still balanced which shows ability of system under unbalanced load conditions. At time instant of 0.21 s the load linked to phase "a" is removed and again it is reconnected at time instant of 0.3 s. At the time of these variations, the internal signals behaviour of the control are also affected. They stabilized themselves at new values for a couple of cycles. This displays the quick responses of the control that has been used.

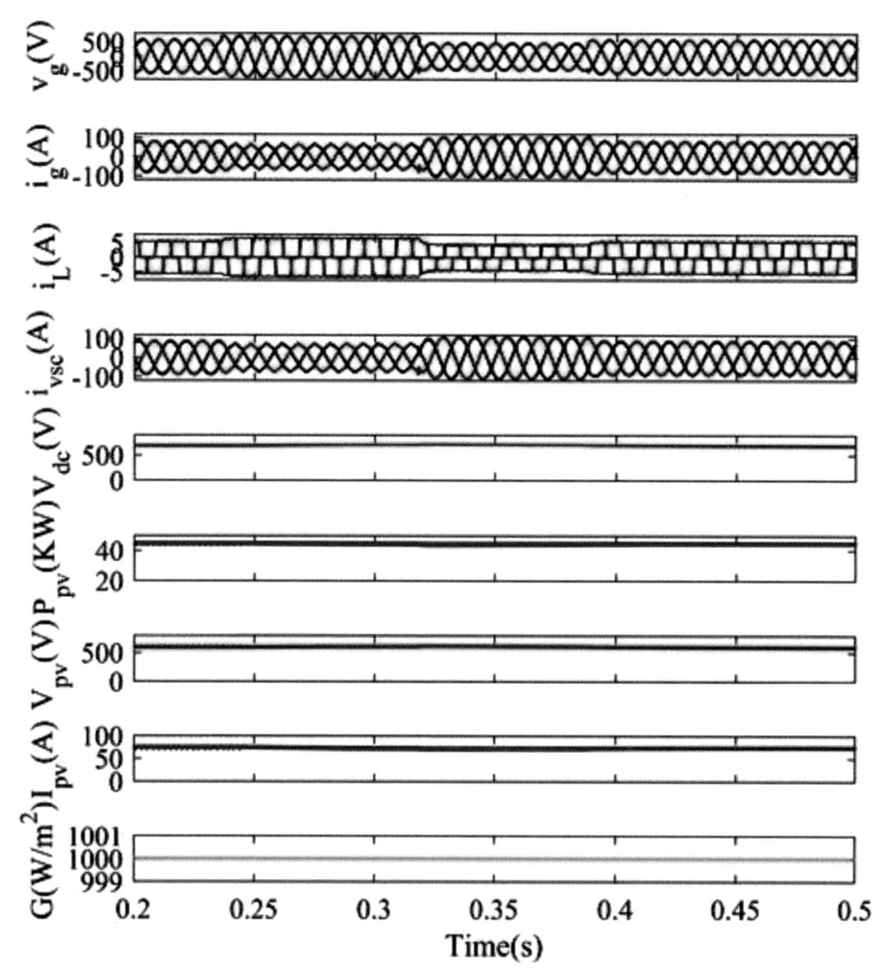

Figure 6.20 Response under grid voltage sag and swell conditions.

6.3.5 Response of system under sag and swell conditions of grid voltage

Figure 6.21 represents the system performance when voltage sag and swell occur in the grid voltage. When sag occurs in the grid voltage, the grid current is increased due to constant PV power supply to the grid. At the same time when swell is noticed in the grid voltage, the grid current gets reduced. It shows that PV supply is unaffected due to a change in grid voltage and current, since it supplies a constant power. This leads to enhanced performance of the system. Figure 6.21 (a–c) show the grid current THD, which is less than 2% and that lies well within IEEE standards. This shows that the system is working efficiently to reduce the harmonics in the grid current. The load current THD is 26.91%.

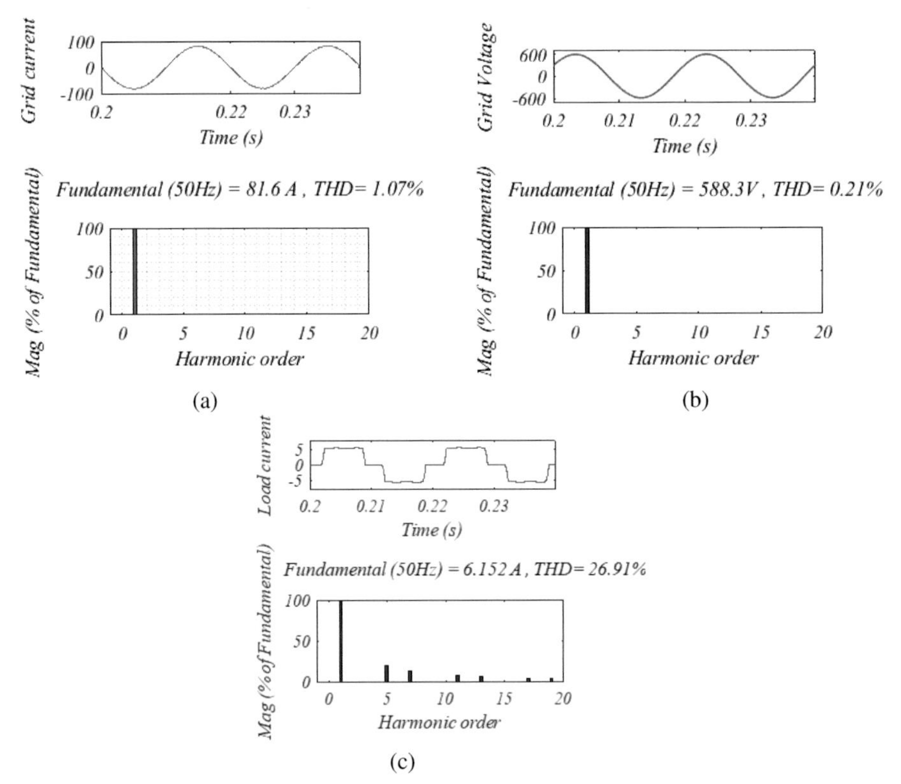

Figure 6.21 THD analysis of grid and load currents and grid voltages.

6.4 Conclusions

EMI/CM levels are decreased in the resultant inverter architecture. Single-phase transformerless inverters that were connected to the grid had their designs examined for differences. All PV inverters' working principle is explained for positive, negative, and zero cycles in details. To enhance selectivity, this research thoroughly examines each topology and weighs the benefits and drawbacks of each. At 2.5 kW, loss calculation is done for several grid-connected transformerless inverter systems. An efficiency analysis is made to see the performance evaluation of HERIC-based modified topology with two well-known topologies named as H5 and HERIC. H5 and HERIC may now process at lower PV voltages because of the addition of a DC-DC converter. Comparing the suggested topology to the HERIC topology, the proposed topology suffers from larger conduction losses. This is due to increase in the number of switches in conduction path. Using partial power instead of full power in the proposed multilevel-based topology at 280 PV voltage offers a significant advantage in terms of overall loss reduction, including switching losses. This advantage is observed to be greater than that of the HERIC plus boost topology. Both the H5 and the HERIC are more efficient at higher voltages (greater than AC peak voltage). Increased switching losses from the new HERIC system account for most of its extra losses. The HERIC + boost and H5 + boost topologies lose efficiency when PV voltage declines. The suggested method, on the other hand, has a substantially lower efficiency loss. With the new inverter design, lower voltage levels are more efficient than the other two topologies. In a modified HERIC-based design, the DC-DC converter only must handle a fraction of the total power. Modified HERIC-based inverter architecture has the following key features: It is more efficient at low insolation than the HERIC + boost and H5 + boost systems since there is no multi-stage power conversion. The MATLAB model has been developed for PV-based DSTATCOM system and it has been controlled using dq control algorithm. Power factor enhancement, harmonic reduction, and DC bus voltage management have been achieved to improve the power quality. The dynamic response and steady state stability of the system have been found satisfactory. The THD in grid current is well within the IEEE standard guidelines.

6.5 Future Works

In the future, understanding power converters will be necessary to integrate renewable energy resources into the grid and to complete the grid demands designated by the grid operator with some harmonic injection. Although low-cost PVs are already accessible, we must have solid data analysis and information about the fabrication materials used in PV panels to ensure optimal efficiency. Furthermore, the widespread availability of thin-film PV panels and wide-bandgap devices will almost certainly lead to new areas of study, displacing the more efficient and widely used translation facilities. The study also can be done to improve energy density and reliability and to improve the overall efficiency of power converters. The use of multilevel inverters to provide a mid-voltage grid connection is another key topic for a transformerless inverter and the future power conditioner Quasi Z Source Network. SiC devices will be primarily employed for converters in the near future, and upgrading GaN converters with SiC devices will enhance the converter's efficiency.

Acknowledgement

The authors would express their gratitude towards IBITF funding agency, India under the SERB, and NAMPET-III, for supporting this work.

References

[1] Khadem, Md. Shafiuzzaman Khan & Basu, Malabika & Conlon, M. (2010). "Power Quality in Grid-connected Renewable Energy Systems: Role of Custom Power Devices." Renewable Energy and Power Quality Journal. 8. 505. 10.24084/repqj08.505

[2] Ellabban, Omar & Abu-Rub, Haitham & Blaabjerg, Frede, 2014. "Renewable energy resources: Current status, future prospects and their enabling technology," Renewable and Sustainable Energy Reviews, Elsevier, vol. 39(C), pages 748-764.

[3] Iea.org/reports/india-energy-outlook-2021

[4] Solar Energy and Wind energy Installed GW Capacity-Global and by Country. Available online: http://www.fipowerweb.com/Renewable-Energy.html (accessed on 20 July 2021

[5] S. Kumar, and B. Singh, "Linear coefficient function based control approach for single stage SPV system integrated to three phase distribution system," IET Gener. Transm. Distrib., 11: 676-684, 2017. https://doi.org/10.1049/iet-gtd.2016.0630

[6] F. Spertino, G. Graditi, Power conditioning units in grid-connected photovoltaic systems: A comparison with different technologies and wide range of powerratings, SolarEnergy, Volume 108, 2014, Pages 219229, ISSN 0038092X, https://doi.org/10.1016/j.solener.2014.07.009. (https://www.sciencedirect.com/science/article/pii/S0038092X14003569).

[7] Q. Chen, "Opportunities and challenges to power electronics industry in alternative and renewable energy," 2009 3rd International Conference on Power Electronics Systems and Applications (PESA), 2009, pp. 1-1.

[8] A. Nami, "Power Electronics for Future Power Grids: Drivers and Challenges," 2018 20th European Conference on Power Electronics and Applications (EPE'18 ECCE Europe), 2018, pp. P.1-P.2.

[9] A. Mittal, K. Janardhan and A. Ojha, "Multilevel inverter-based Grid Connected Solar Photovoltaic System with Power Flow Control," 2021 International Conference on Sustainable Energy and Future Electric Transportation (SEFET), 2021, pp. 1-6, doi:10.1109/SeFet48154.2021.9375753.

[10] M. Azri and N. A. Rahim, "Transformerless power converter for grid-connected PV system with no-ripple input current and low ground-leakage current," 3rd IET International Conference on Clean Energy and Technology (CEAT) 2014, 2014, pp. 1-6, doi:10.1049/cp.2014.1503.

[11] Tan, Freddy & Abd Rahim, Nasrudin & Hew, Wooi. (2013). Three-phase transformerless grid-connected photovoltaic inverter to reduce leakage currents. 277-280. 10.1109/CEAT.2013.6775640.

[12] G. Vazquez, P. R. Martinez-Rodriguez, J. M. Sosa, G. Escobar, M. A. Juarez, and A. A. Valdez, "H5-HERIC based transformerless multilevel inverter for single-phase grid-connected PV systems," IECON 2015 - 41st Annual Conference of the IEEE Industrial Electronics Society, 2015, pp. 001026-001031, doi:10.1109/IECON.2015.7392234.

[13] M. S. B. Ranjana, N. Sreeramula Reddy and R. K. P. Kumar, "A novel single-phase advanced multilevel inverter with adjustable amplitude of voltage levels," 2014 International Conference on Circuits, Power and Computing Technologies [ICCPCT-2014], 2014, pp. 950-957, doi:10.1109/ICCPCT.2014.7054950.

[14] A. Anurag, Y. Yang and F. Blaabjerg, "Reliability analysis of single-phase PV inverters with reactive power injection at night considering mission profiles," 2015 IEEE Energy Conversion Congress and Exposition (ECCE), 2015, pp. 2132-2139, doi:10.1109/ECCE.2015.7309961.

[15] Yu, Wensong et al. "High-Efficiency MOSFET Inverter with H6-Type Configuration for Photovoltaic Nonisolated AC-Module Applications." IEEE Transactions on Power Electronics 26 (2011): 1253-1260.

[16] V. L. Srinivas, S. Kumar, B. Singh and S. Mishra, "A Normalized Adaptive Filter for Enhanced Optimal Operation of Grid-Interfaced PV System," IEEE Transactions on Industry Applications, vol. 57, no. 2, pp. 1715-1724, March-April 2021.

[17] Jedtberg, H., Pigazo, A., Liserre, M., & Buticchi, G. (2017). "Analysis of the Robustness of Transformerless PV Inverter Topologies to the Choice

of Power Devices," IEEE Transactions on Power Electronics, 32, 5248-5257.

[18] H. Abubakr, J. C. Vasquez, K. Mahmoud, M. M. F. Darwish, and J. M. Guerrero, "Comprehensive Review on Renewable Energy Sources in Egypt— Current Status, Grid Codes, and Future Vision," in IEEE Access, vol. 10, pp. 4081-4101, 2022, DOI:10.1109/ACCESS.2022.3140385

[19] S. Kumar and B. Singh, "Implementation of Grid Interactive and Islanded Controlled PV System for Residential Applications," 2020 IEEE Industry Applications Society Annual Meeting, Detroit, MI, USA, 2020, pp. 1-6

[20] Vb Calais, M.; Agelidis, V.G.; "Meinhardt, M. Multilevel converters for single-phase grid connected photovoltaic systems: An overview." Sol. Energy 1999, 66, 325–335.

[21] Tuan Ngo and Surya Santoso, "Grid-connected Photovoltaic converters: Topology and grid interconnection", Journal of Renewable And Sustainable Energy 6,032901 (2014) https://doi.org/10.1063/1.4876415.

[22] Xu Zheng, Wang Hui and Li Youchun, "A hybrid modulation scheme for grid-connected photovoltaic systems," 2008 Third International Conference on Electric Utility Deregulation and Restructuring and Power Technologies, 2008, pp. 2707-2711, DOI:10.1109/DRPT.2008.4523870.

[23] G. Vazquez, P. R. Martinez-Rodriguez, J. M. Sosa, G. Escobar, M. A. Juarez, and A. A. Valdez, "H5-HERIC based transformerless multilevel inverter for single-phase grid connected PV systems," Proc. IEEE Ind. Electron. Soc. Annu. Conf., Nov. 2015, pp. 001 026–001 031

[24] Li, W.; Gu, Y.; Luo, H.; Cui, W.; He, X.; Xia, C., "Topology review and derivation methodology of single-phase transformerless photovoltaic inverters for leakage current suppression." IEEE Trans. Ind. Electron. 2015, 62, 4537–455

[25] Zhang, L.; Sun, K.; Xing, Y.; Zhao, J., "Parallel Operation of Modular Single-Phase Transformerless Grid-Tied PV Inverters with Common DC Bus and AC Bus." IEEE J. Emerg. Sel. Top. Power Electron. 2015, 3, 858–869.

[26] S. Adam, V. Hahanov, S. Chumachenko, E. Litvinova and K. Man, "Novel Design and Simulation of HERIC Transformerless PV Inverter in MATLAB/Simulink," 2021 IEEE East-West Design & Test Symposium (EWDTS), 2021, pp. 1-4, doi:10.1109/EWDTS52692.2021.9581017.

[27] Y. Huang, X. Yuan, J. Hu, P. Zhou and D. Wang, "DC-Bus Voltage Control Stability Affected by AC-Bus Voltage Control in VSCs Connected to Weak AC Grids," IEEE Journal of Emerging and Selected Topics in Power Electronics, vol. 4, no. 2, pp. 445-458, June 2016, doi:10.1109/JESTPE.2015.2480859.

[28] Deepa R, A. Nagashree and K. Anand, "Transformerless grid-connected single-phase inverter for power system application," 2015 International Conference on Emerging Research in Electronics, Computer Science and Technology (ICERECT), 2015, pp. 418-422, DOI:10.1109/ERECT.2015.7499052.

[29] Y. P. Siwakoti and F. Blaabjerg, "H-Bridge transformerless inverter with common ground for single-phase solar-photovoltaic system," 2017 IEEE Applied Power Electronics Conference and Exposition (APEC), 2017, pp.2610-2614, doi:10.1109/APEC.2017.7931066.

[30] A. F. de Sousa Silva, S. Pires Pimentel, E. G. Marra and B. Pinheiro de Alvarenga, "Energy-Balance Based Voltage Regulation Method for Multiple DC-links in Asymmetrical Cascaded Multilevel Inverters," 2019 IEEE 15th Brazilian Power Electronics Conference and 5th IEEE Southern Power Electronics Conference (COBEP/SPEC), 2019, pp. 1-6, DOI:10.1109/COBEP/SPEC44138.2019.9065772.

[31] Wang, J.; Ji, B.; Zhao, J.; Yu, J., "From H4, H5 to H6 Standardization of full-bridge single-phase photovoltaic inverter topologies without ground leakage current issue." In Proceedings of the 2012 IEEE Energy Conversion Congress and Exposition, Raleigh, NC, USA, 15–20 September 2012; 2419–2425.

[32] A. Ashraf Gandomi, K. Varesi and S. H. Hosseini, "DC-AC buck and buck-boost inverters for renewable energy applications," The 6th Power Electronics, Drive Systems & Technologies Conference (PEDSTC2015), 2015, pp. 77-82, DOI:10.1109/PEDSTC.2015.7093253.

[33] M. H. Ghaderi, A. Sarikhani and M. Hamzeh, "A New Transformerless Common-Ground Single-Phase Inverter for Photovoltaic Systems," 2019 10th International Power Electronics, Drive Systems and Technologies Conference (PEDSTC), 2019, pp. 259-264, DOI:10.1109/PEDSTC.2019.8697828.

[34] G. Vazquez, P. R. Martinez-Rodriguez, J. M. Sosa, G. Escobar, M. A. Juarez and A. A. Valdez, "H5-HERIC based transformerless multilevel inverter for single-phase grid-connected PV systems," IECON 2015 - 41st Annual Conference of the IEEE Industrial Electronics Society, 2015, pp. 001026-001031, doi: 10.1109/IECON.2015.7392234.

[35] Application Note SMA: "Capacitive Leakage Current," Jan. 2009. [Online]. Available: http://files.sma.de/dl/7418/Ableitstrom-TI-en-25.pdf

[36] A. Saha *et al.*, "Comparative Study of Different Transformer-less Inverter Topologies for Grid-tied Photovoltaic System," 2019 5th International Conference on Advances in Electrical Engineering (ICAEE), 2019, pp. 783-788, doi:10.1109/ICAEE48663.2019.8975675.

[37] P. Chamarthi, M. Moursi, V. Khadkikar, K. Hosani, and T. Fouly, "A Novel Single-Phase Voltage Boosting Transformer Less Inverter Topology for Grid-connected Solar PV Application," 2021 International Conference on Sustainable Energy and Future Electric Transportation (SEFET), 2021, pp. 1-6, DOI:10.1109/SeFet48154.2021.9375739.

7

HVDC System Issues and Related Solutions using Soft Computing

Ravi Shankar Tiwari[1,2], Om Hari Gupta[1], Vijay K. Sood[3]

[1]NIT Jamshedpur, India
[2]GLA University, India
[3]Ontario Tech University, Canada
Email: 2018rsee017@nitjst.ac.in, omhari.ee@nitjsr.ac.in,
ravishankar.tiwari@gla.ac.in2, vijay.sood@ontariotechu.ca

Abstract

HVDC systems are now extensively used in power transmission networks worldwide. Due to fast, precise, and flexible control of DC link current, HVDC systems play a significant role in modern power systems and smart grids. HVDC systems are classified into different types based on application, modelling, and control features. The LCC-HVDC system is basically used for bulk power transmission and VSC-HVDC is preferred for integration of renewable energy sources (RESs) in a smart grid. Thus, to achieve successful operation of HVDC links with RES, extensive planning and research should be performed before installing HVDC to integrate renewable sources. An incorrect sizing of the HVDC converter may lead to unnecessary curtailment of renewables or underutilization of converters. The provisions for a virtual generator and load shedding options are highly recommended in a power system with extensive RES penetration. These provisions are necessary because the modern grid consists of large percentage of RES; this reduces the overall inertia of power system. Hence, a sudden heavy change in load may cause the frequency instability. Thus, together with the promising features of HVDC system there are number of other issues that also arise due to evolving changes in a power system. Some of these are issues related to optimization and cost-effectiveness in operation, electromagnetic interference, security, day ahead operational strategies to maximize the accommodation of renewable energies and challenges related to fault detection and diagnosis. The low cost, traceable, and simple solutions to such issues can effectively be provided by using soft computing (SC) techniques. This chapter summarizes the results of a few SC methods applied in HVDC systems to resolve various operational, control, and protection problems. Finally, real-time simulation results and hardware-in-loop (HIL) model of a VSC-HVDC test system is presented.

Keywords: HVDC transmission, line commutated converter (LCC), voltage source converter (VSI), soft computing, smart grid, artificial neural network, fuzzy logic, and wavelet analysis

7.1 Introduction

HVDC systems are one of the alternate choices for energy transmission over long distances and to integrate renewable generation into conventional grids. The HVDC system can provide solutions to various issues related to power quality, power system stability, economy, and power flow control with associated HVAC transmission systems. The HVDC system based on voltage source converters (VSC) is getting popular due to its capability to address several issues as compared to the current source converter (CSC). However, there are several issues that can affect the operation and control performance of either VSC or CSC HVDC systems. The sag in terminal voltage on the inverter side due to an increase in power demand affects the operation of the HVDC system. The intermittent renewable source connected to the HVDC grid becomes a significant issue that affects grid operation. New challenges in achieving fault ride-through of HVDC and renewable generation arise due to the complex interface between the power grid and renewable generations. The combined fault ride-through realization in HVDC and the sending or receiving terminals of renewable energy sources increase system availability with safe and stable operation [1]. Integrated renewable sources introduce the dynamic aspects of the power system, such as transient and small-signal stability (SSS). Thus, it is crucial to investigate the relation between frequency control and system response to a disturbance. Investigating the probable side-effects of such a system, in the frame of electromechanical dynamics, is also important. Thus, proficient coordination of control elements is required to mitigate such stability issues in the power systems. The asynchronous HVDC interconnection is a perceived solution for such stability issues due to its high-speed flexible control behaviour. Although the HVDC system offers a wide variety of control modes, such a robust and straightforward solution becomes more attractive for a system operating under high uncertainty. The active power regulation of the HVDC proportional to change in frequency is often considered a safe alternative for ensuring frequency stability. However, certain case studies have discovered scenarios in which the frequency support supplied by an HVDC system has a negative impact on the SSS margin. This phenomenon appears due to the high gain of power controller supporting frequency stability due to deterioration in system damping. Therefore, an optimal gain of the HVDC power controller, concerning the damping coefficient and the positive impact of the HVDC controller on synchronizing coefficients, is an important issue in VSC-HVDC systems [2]. The life expectancy of HVDC control and protection equipment is another issue in the HVDC system. Specifying and monitoring desirable factors or designs mentioned in the project commencement stage facilitates the replacement of the control system as required. Plant outage risk assessment, economic evaluation of plant unavailability, refurbishment duration, source code and programming tool to modify control, etc., are some factor-refurbishments [3]. The closed-loop testing of the HVDC system in a real-time digital simulator is used to validate the CSC valve group, multi-modular converter, frequency-dependent DC line, and switchable filter models. The issues present with the real-time digital simulation for HVDC systems, its implementation, background, application, and model validation are addressed in [4].

7.2 HVDC Systems and Their Significance in a Smart Grid

The large participation of renewable sources also needs a higher reserve margin for power balancing during an emergency and flexible power flow control. Power system planning and transmission operators must consider other challenges such as asynchronous interconnection, charging currents, environmental impact due to new transmission corridor, cost, efficiency target, and global de-carbonization.

The high voltage direct current (HVDC) transmission can address the above challenges, due to its inherent advantage in long-distance transmission and flexible control ability. It is extensively used for long-distance bulk power transmission, asynchronous grid interconnections, stability improvement of AC system, power flow control, long under-sea transmission, and renewable energy integration. Another advantage of the HVDC system is the reduction in right of way (ROW) [5]. The HVDC transmission consists of converter stations, DC transmission lines, harmonic and ripple filters, converter transformers, smoothing reactors, etc. The sending-end converter acts as a rectifier and receiving-end converter acts as an inverter. The converters significantly impact the efficiency and rated power capacity of HVDC transmission. A properly dimensioned converter decreases harmonics, improves power transmission capabilities, and promotes reliability by providing high fault tolerance along the line [6]. Around the world, various HVDC converter topologies have been used. The two most common forms are the line-commutated converter (LCC) and voltage source converter (VSC).

LCCs are built using thyristor technology, which consists of semiconductor devices featuring four layers of P and N-type material, functioning as bi-stable switches. They are activated by a gate pulse and stay in the "on" state until the subsequent current zero-crossing. LCC converters require a highly robust synchronous voltage source to reliably facilitate commutation, making them unsuitable for black start operations. Among all HVDC converter technologies, LCC boasts the highest voltage and power rating capacity, supporting current ratings of up to 6250 A and blocking voltages of 10 kV [7–9].

LCC provides control on both the rectifier and inverting sides by adjusting the firing angle α. It employs a unidirectional line commutated flow of DC power that is injected into a receiving AC network. In an LCC, the output DC voltage is maintained at a constant level; thus, it is referred to as a current source converter (CSC).

Within an LCC system, the current direction remains unchanged when power is reversed between stations; this reversal is achieved by merely flipping the DC voltage polarity at both ends. This method is exceptionally dependable and demands minimal maintenance. It stands as the most efficient means for transmitting large-scale power over high-voltage transmission lines. Due to these attributes, LCC technology stands as the most commonly employed HVDC scheme [10].

On the other side, the VSC is based on insulated gate bipolar transistor (IGBT) or gate turn-off (GTO) technologies. Its turn-on and turn-off instants are both controlled independently irrespective of AC voltage magnitude; thus, it is suitable for a black start operation [11]. A VSC-HVDC system provides the benefits of independent active and reactive power control, faster response, and ability to connect to a weak AC system. Thus VSC-HVDC is a popular choice for the integration and planning of a modern smart grid containing a multi-terminal system or DC grid. Conversely, VSC-HVDC systems come with a higher price tag and increased energy losses stemming from their high-frequency switching, unlike LCC-HVDC systems. Additionally, the VSC system has lower voltage and power ratings, primarily because of device limitations. A potential remedy for addressing the constraints of the VSC-HVDC system involves the implementation of a hybrid HVDC system, which amalgamates the benefits offered by both LCC and VSC-HVDC systems [12]. The LCC system cuts down on costs, while the VSC system mitigates issues related to commutation failures.

The HVDC system can be configured using various topologies, depending on the number of overhead lines (OHLs) or cable conductors used, length of cable or OHL, based on converter technology, and integration points of the AC grid. Figure 7.1 and Table 7.1 show the basic classification of various HVDC configurations based on application.

Table 7.1 Classification of HVDC link based on applications.

Sr. No.	HVDC topology	Remarkable feature	Applications
1.	Back-to-back HVDC system	• Rectifier and inverter stations installed at same location • Integrate asynchronously operating grids	In meshed grid to achieve a defined power flow
2.	Monopolar HVDC system	• Bulk power transmitted to long-distance • Overhead conductor (OHC) operates at negative polarity • Ground, sea, or metallic return paths act as forward path	Applied for river or sea crossing Power supply to island
3.	Bipolar HVDC system	• For transmitting bulk power • Two independent OHCs are used as positive and negative pole • Ground or metallic return • Operates as MP-HVDC with 50% capacity during maintenance or fault on one pole	For distance greater than 500 km. Cost-effective, reliable, and lower losses Improve stability with by regulating power flow in parallel AC line.
4.	Multi-terminal HVDC system	• More than two sets of converters operating as a rectifier or inverter • VSCs are preferred to construct MT-HVDC	Used to integrate RES and to develop DC super-grid Monopole or bipolar operation achieved by switching out a few converters
5.	Hybrid HVDC	• LCC is used at rectifier and VSC at the inverter station	Black start capability

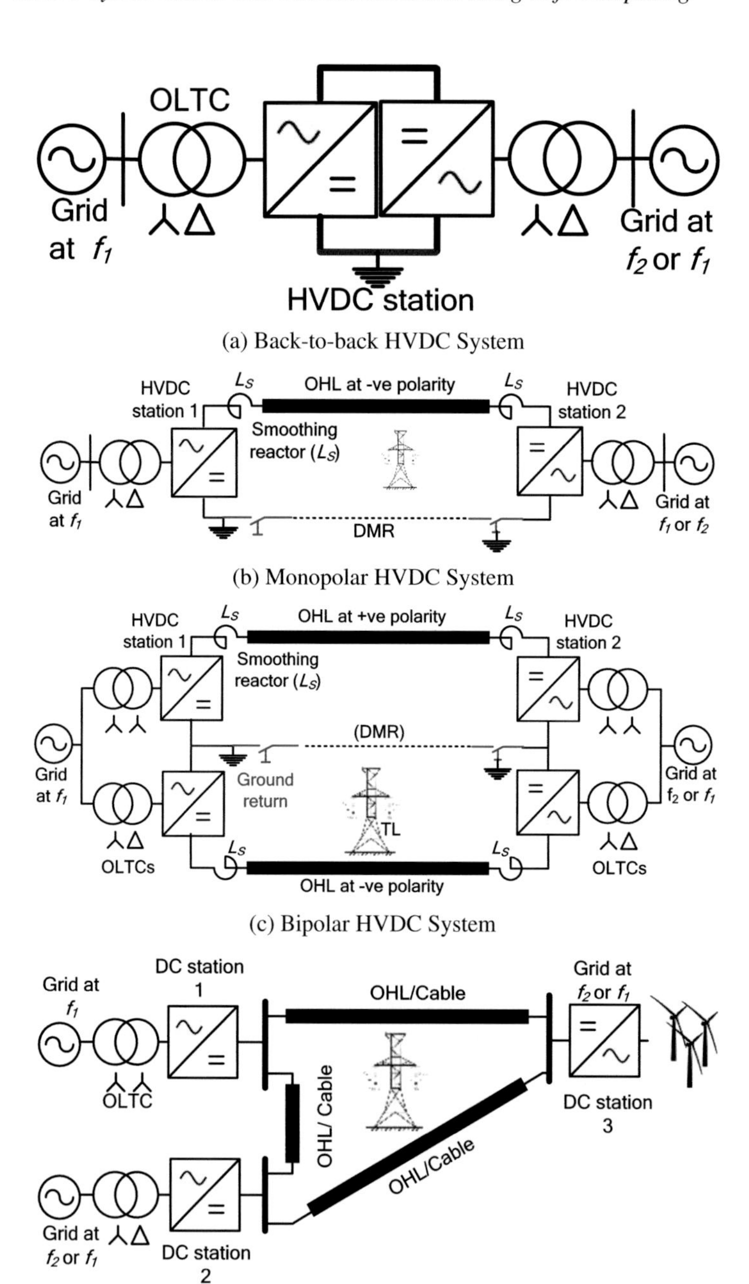

(a) Back-to-back HVDC System

(b) Monopolar HVDC System

(c) Bipolar HVDC System

(d) MT-HVDC System

Figure 7.1 HVDC system configurations [11, 12].

7.2.1 HVDC modelling and control

Irrespective of configuration, the major technologies used for HVDC systems are: (a) LCC-HVDC, (b) VSC-HVDC system, and (c) a combination of LCC and VSC devices. The LCC technology consists of a thyristor valve type, 6-pulse power electronic converters as the basic building block of the HVDC system. The converter output voltage is controlled by changing the delay in firing angle α of the thyristor valves. The firing angle α varies from 0° to 90° for rectifier and 90° to 180° for inverter mode of converter operation. The mean converter output DC voltage (V_{dc}), line current (I_{dc}), and power flow at rectifier (P_1) and inverter stations (P_2) of LCC-HVDC transmission are given by eqn (7.1), (7.2), and (7.3), respectively. Its schematic diagram with the control arrangement is shown in Figure 7.2.

$$V_{dc} = \frac{3\sqrt{2}}{\pi} V_L \cos\alpha, \tag{7.1}$$

$$I_{dc} = \frac{V_{dr} - V_{di}}{R_L}, \tag{7.2}$$

$$P_1 = V_{dr} I_{dc} \quad or \qquad P_2 = V_{di}(-I_{dc}). \tag{7.3}$$

The principal control characteristics of LCC-HVDC system are shown in Figure 7.3 and the main control components are incorporated as follows:

a. Constant current (CC) – represented by BC line segment at rectifier
b. Constant current (CC) – represented by HG line segment at inverter
c. Constant extinction angle (CEA) – represented by IJ segment at inverter
d. Current error – represented by HI line segment at inverter

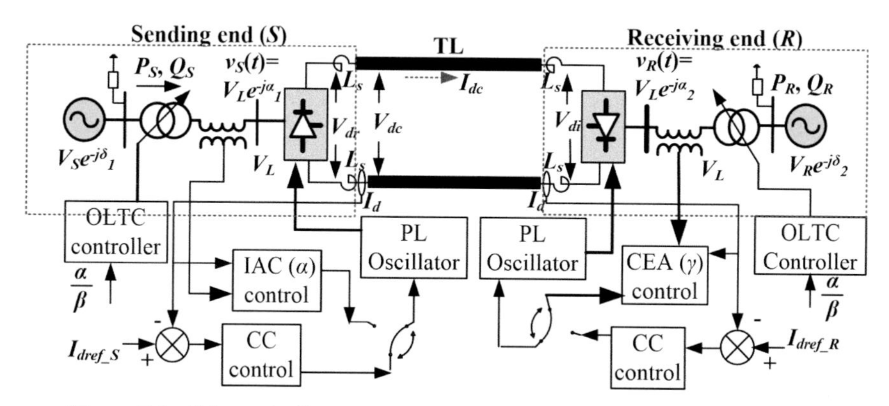

Figure 7.2 Schematic diagram and control arrangement of LCC-HVDC system.

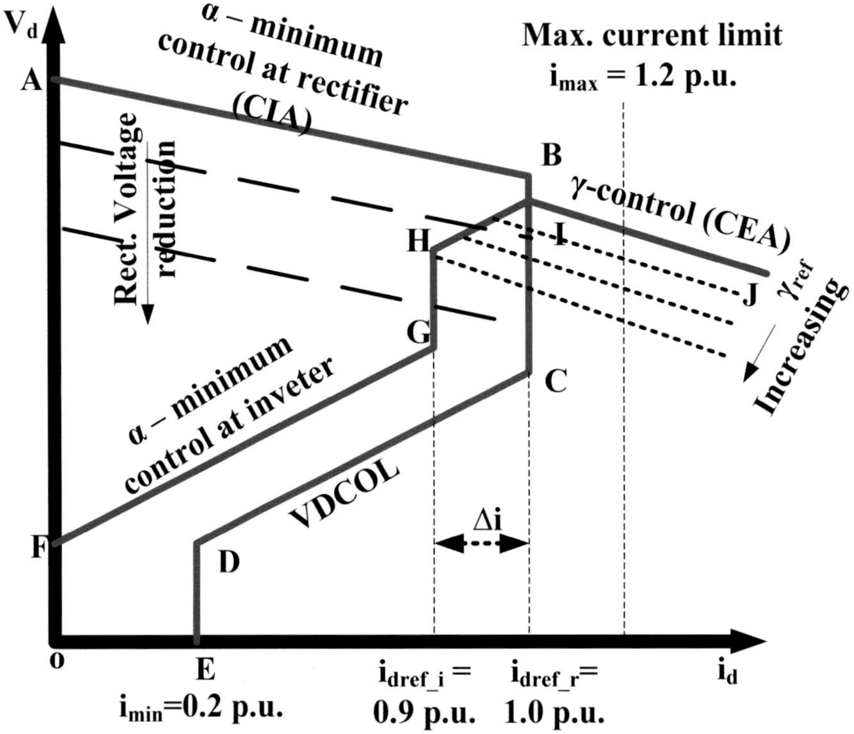

Figure 7.3 LCC-HVDC control characteristics.

e. Minimum ignition angle – at rectifier represented by AB line segment
f. Voltage-dependent current order limit (VDCOL) – represented by DC and GF segments at rectifier and inverter, respectively.

The meeting point of CC and CEA characteristic is a point of normal operation at which the HVDC systems operates. Operating HVDC at intersecting point reduces demand of reactive power and least chance of commutation failure. During operation as voltage at rectifier decreases, the rectifier starts operating in CIA and inverter at CC characteristic.

The VSC-HVDC technologies are based on insulated gate bipolar transistor (IGBT) devices, and have capabilities for STATCOM, black start, feed ability to weak AC networks, and power flow reversal without changing polarities. The VSC-HVDC normal operation and control modes is shown in Figure 7.4 and are governed by eqn (7.4), (7.5), and (7.6) [13] as given below.

$$P_{s1} = 3\frac{V_{s1}V_{c1}}{\omega L}\sin(\delta_1 - \alpha_1), \tag{7.4}$$

$$Q_{s1} = 3\frac{V_{s1}(V_{s1} - V_{c1}\cos(\delta_1 - \alpha_1))}{\omega L}, \tag{7.5}$$

$$U_{d1} = \frac{\sqrt{2}V_{c1}}{m_1}, \ (0 \leq m_1 \leq 1), \tag{7.6}$$

where P_{s1} and Q_{s1} are the active and reactive powers respectively, provided by the AC system, and P_{c1} and Q_{c1} are the active and reactive powers respectively, absorbed by the VSC. The V_{s1}, V_{c1}, U_{d1}, and $\delta_1 - \alpha_1$ are the root mean square (rms) voltage at the AC system, input to the VSC converter, DC output voltage, and phase angle between V_{s1} and V_{c1}, respectively. The m_1 is the modulation index for the pulse width modulation (PWM) when the utilization factor is unit. The converter controls an appropriate magnitude (V_{c1}) and a phase angle of AC voltage to cause the flow of reactive and active powers in the required direction at the sending and receiving ends of the HVDC system. In which, one converter will maintain the constant DC voltage while the other adjusts it to the current flow (and hence active power flow) in the DC line.

A VSC-HVDC configuration and control modes are shown in Figure 7.4. The typical control modes of VSC-HVDC are given as follows:

a) **Mode I:** VSC operates in either AC voltage control or reactive power control mode together with voltage control of DC side.

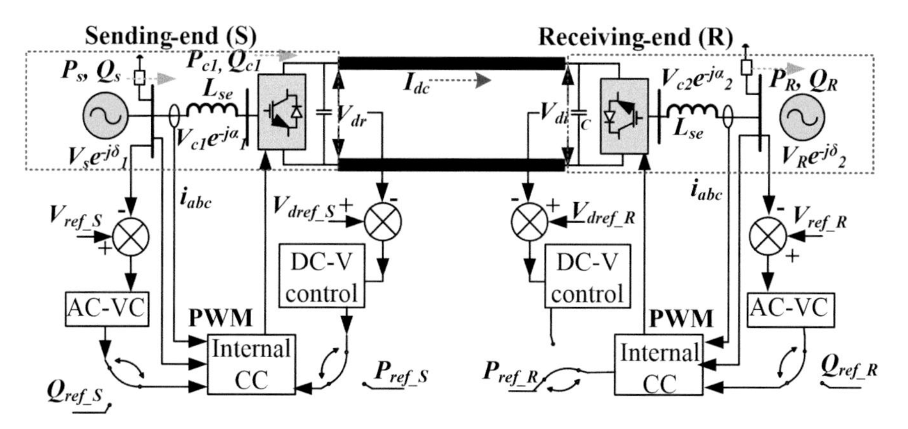

Figure 7.4 Schematic diagram and control arrangement of VSC-HVDC system.

b) **Mode II:** VSC operates in real power control or voltage (V_{dc}) control mode i.e., if one converter operates in active power control then the other will be in DC voltage control mode together with mode I.
c) **Mode III**: VSC operates for controlling power factor and frequency control.

7.2.2 Significance of HVDC in smart grid

The electrical network with features to integrate renewable sources, bidirectional power flow, preventive power outage, smart metering, and bidirectional communication of information, clean energy sustainability, and efficiency is called smart grid technology. Moreover, the cybersecurity, controllability, and measurability of energy at each point are unavoidable in the modern power grid [14], [15]. The smart grid's other features are stability, controllability, accuracy, economy, optimality, reliability, availability, and flexibility. The power grid becomes robust enough against power failures, voltage sag, losses, frequency, voltage fluctuation, over-current, and over-voltages with the implementation of the above-mentioned features of the smart grid [16]. Unlike the conventional grid, a consumer can be a prosumer in a smart grid by selling excess renewable energy to the grid. Thus, the smart grid brings the policymakers, network operator, consumers, and prosumers on a common platform to fulfill the combined goal of efficient, reliable, and economical availability of electrical energy.

The classical power grid undergoes a paradigm shift in transmission and distribution networks to incorporate features of smart grids. Installation and commissioning of HVDC systems for integrating RES and bulk power transportation is one of those features of smart grid [17]. Lower stability limit, unfeasible operations with asynchronous grids, integration of renewable energy, and large reactive compensation are the main drawbacks of EHV AC transmission systems. However, the smart grid initiative focuses on minimizing these shortcomings.

At multiple instants and situations, HVDC transmission provides better flexibility and functionality for the smart grid as compared to AC transmission. Examples of such challenges are uncertainty in the amount and location of renewable energy, weak transmission networks for energy evacuation up to the load centres and offshore or interconnections involving water crossings, reactive power compensation, frequency, and voltage control to mention a few. The various configurations of the HVDC system that give a solution for these challenges in smart grid projects are:

7.2.2.1 Improving transient stability

Due to precise, fast, and fully controllable power flow, HVDC systems can improve transient stability of asynchronous or synchronous interconnected power systems. Authors in [18] carried out an investigation to improve the transient stability of the AC network by modulating the reactive and real powers of the VSC-HVDC link. Normally, the energy dispatch and control centre decide the amount of power to be delivered over the DC line, and the outstanding capacity of the converter is to maintain magnitude of voltage on the AC bus. Additionally, in VSC-based HVDC systems, oscillations in power network are damped by modulating real and reactive powers exchange. Modulating active power increases damping oscillation by four times than reactive power modulation. In [19], authors compare the effectiveness of oscillation damping using VSC and LCC-HVDC systems. Figure 7.5 shows power flow, line current, voltage, and power angle after an AC line fault in the hybrid system interconnection. Results show that power flow control by HVDC dampens the oscillation and restores pre-fault system conditions quickly. Without HVDC damping, the oscillation increasingly leads to permanent tripping of the interconnected system. In [20], authors present a

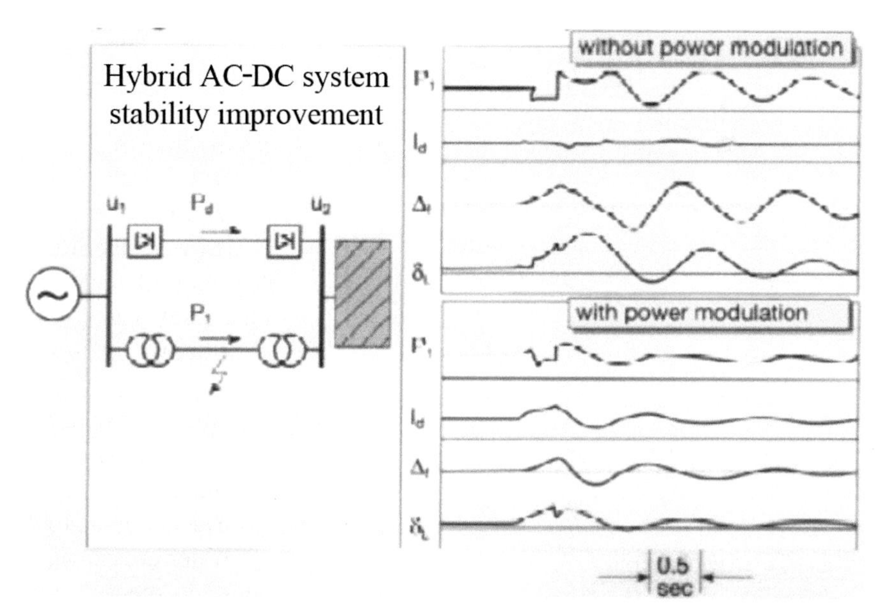

Figure 7.5 Hybrid AC-DC system stability improvement during single-phase line-to-ground fault.

cooperative active and reactive powers control of HVDC system to stabilize power system including wind farms.

7.2.2.2 Flexibility

In deregulated trading markets with significant presence of renewables, where operators can appreciate the importance of new ancillary services to handle fluctuation in power flows, there is an opportunity to integrate the highly flexible VSC-HVDC systems directly as part of the smart grid structural design. Given that there is notable development in VSCs' the voltage and current capacity have greatly improved in recent times. Additional functionality might be coordinated carefully for most power markets matching with the real power importing and exporting capability of converters.

7.2.2.3 Power restoration

The VSC-HVDC is most suitable for power restoration in thermal power plants during black start or major outage conditions. Since the outage can affect any one of the two AC systems connected via HVDC line. Thus, the DC line can be worn to supply cranking power to startup the auxiliaries of large thermal power plants to be started. Moreover, the DC line helps to ramp up the minimum power necessary to avoid load rejection after starting and successfully continue operation. The droop control with local frequency integrator at the rectifier station, it is possible to deploy a suitable restoration procedure. A demonstration of entire restoration path together with a thermal unit, DC line, an AC OHL as well as local load is simulated [21].

7.2.2.4 Economic

When compared to HVAC systems, the right of way (ROW) of HVDC systems is narrower in terms of environmental and societal considerations. For a given transmission corridor size, HVDC systems can transport more power than AC systems. Figure 7.6 depicts the ROW comparison of AC and HVDC systems for a transmission capacity of 3000 MW [22]. At places with limited space, it signifies that the better option is to replace the HVAC line into HVDC lines in near future. The Ultranet project in Germany, Hudson project in New York City, and Transbay cable project in San Francisco are few working projects of such HVDC interconnections. Furthermore, lengthy distance high-power HVDC systems with underground or submarine cables are economically and technically possible, whereas charging current limits the behaviour of systems with AC cables.

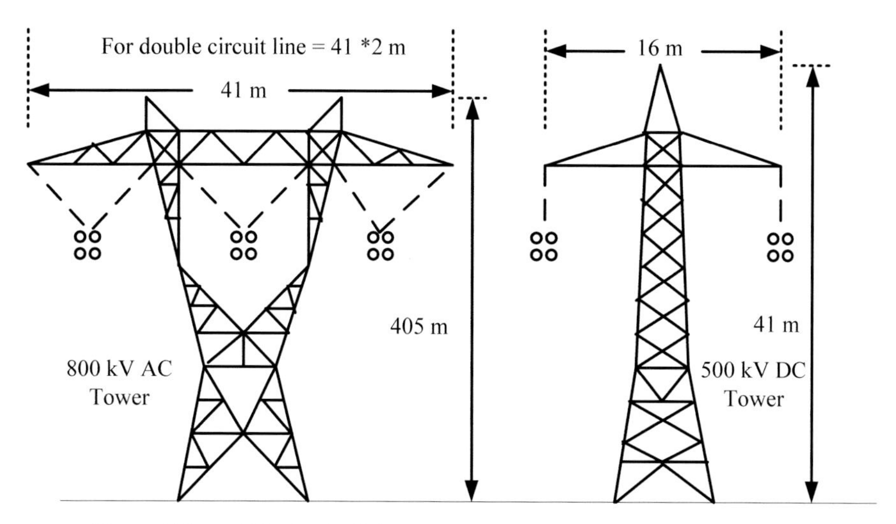

Figure 7.6 ROW of 800 kV AC and 500 kV HVDC for 3000 MW capacity.

7.3 Issues in HVDC Systems Due to Ongoing Modifications in Power System

HVDC systems were introduced to overcome the limitations of traditional HVAC transmission systems such as losses, power flow control, asynchronous interconnection, long-distance transmission, etc. The classic LCC-HVDC technology based on semi-controlled devices, successfully operated with a traditional power system. The operational benefits achieved with LCC-HVDC are lower transmission cost over a long distance, reduced ROW, improved stability, higher loading limits, integrated renewable sources, etc. Traditionally, the classic HVDC economically and efficiently evacuates bulk power from remote generation.

However, the modern power system is facing a paradigm shift from the traditional one. A huge increase in power demand and faster depletion of fossil fuel promotes the installation of large-scale renewable power plants. With the increasing expansion of renewable energy, such as solar photovoltaic (PV) and wind parks, there is substantial demand to accommodate electricity and provide flexible operation by utilizing classic DC transmission technology with semiconductor-controlled devices. The migrated wind farms from coast to offshore, electric vehicles, distributed generations, battery storage, and solar PV system made integration a challenging task. Thus, the major issues can broadly be classified as (a) integration issues, (b) operation and control strategies, (c) optimization and cost-effectiveness, and (d) security issues [23].

7.3.1 Integration issues

The integration issues can be further subdivided as technical issues, issued related to protection, modelling, economic and social, climate, and actual installation and commissioning. The larger deployment of HVDC in existing HVAC systems constitutes a hybrid AC-DC super-grid. Due to their high cost and short time rating commercial DC breakers are not easily accessible. During protection, the complete HVDC grid is tripped for faults in one part of the HVDC system which is not appropriate and inconvenient for the MT-HVDC system. The HVDC system recovery after fault depends on the strength of interconnecting AC grids. The ongoing power system modifications may lead to a change in short circuit ratio (SCR) and thus, the strength of the AC network. The HVDC system operating with a weak AC network of SCR less than 2 is more prone to repetitive commutation failure and transient over voltage for small disturbances. Therefore, proper compliance with grid codes must be available for addressing the operational and integration issues of the hybrid AC-DC network.

7.3.2 Low power quality and electromagnetic interference (EMI) [24]

At the point of common coupling, the power flow is controlled by HVDC converters. Despite several benefits, converters cause a negative impact on the power quality of the electrical system connected on both ends of the HVDC link. The HVDC system consists of complex elements like cables and overhead lines of different resonance characteristics. Its time constant is of the same order as that of the converter connecting the grid. The interaction among them can lead to objectionable behaviour; therefore, investigations are required to assure a safe connection of different elements within the HVDC grid. In [25], the authors discuss the resonance behaviour of the HVDC line and shows the effect of cable length on the stability of power transferred. The increase in power transfer becomes unstable with the reduction of cable length as shown in Figure 7.7.

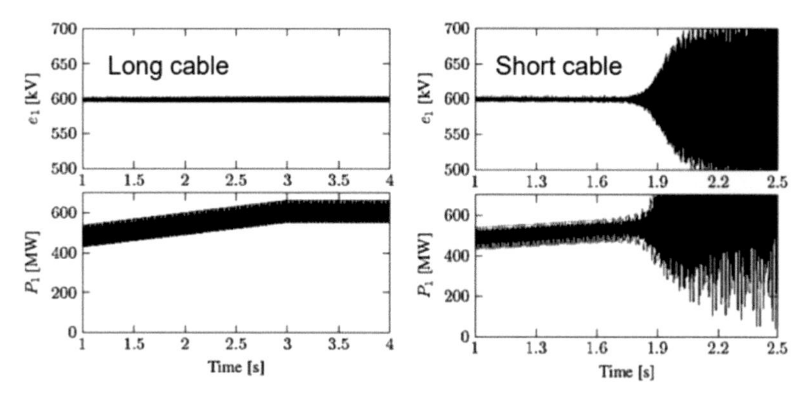

Figure 7.7 Effect of cable resonance behaviour with change in cable length from 50 km to 25 km.

7.3.3 Day-ahead operational strategy of HVDC to maximize accommodation of renewable energy between peak-valley load

The problem of uneven distribution of renewable energy, especially winds from load centres are significantly resolved by providing large-scale transportation facilities via HVDC transmission. The HVDC transmission has the advantage of larger capacity and control over a long distance. It can tap coordinated optimized potential of cross-regional power and accommodate improved renewable energy by flexible control. However, the traditional two-stage fixed power mode of HVDC operation is based on the difference in peak-valley load. It cannot deal with fluctuation in wind power and frequency deviation of HVDC reactive control devices. In [26], the authors proposed a day-ahead forecasting plan for the HVDC tie-line to consider employing additional resources to regulate the reactive power and a coordinated optimization of hydrothermal power at the receiving-end to determine the optimum base power for HVDC transmission. The proposed strategy was a coordinated multi-scale operation to get regulation and base power of HVDC at diverse time scales. This resolves the problem of huge fluctuation in wind power and avoids frequency deviation of HVDC control devices.

7.3.4 Operation and control strategies

To accommodate the evolving modification of power grid, the HVDC systems control strategies, voltage regulation and operating mechanisms must change suitably. The converters in HVDC systems are provided with a wide range of current limits, voltage, and mode of operation. A VSC-HVDC system has three operational modes; these are AC voltage, active power, frequency, and power factor control modes.

7.3.5 Security issues

The HVDC system is most vulnerable to cyber threats due to high digitalization of power system and smart transmission. It is critical to ensure that communicated information and control data must be secure for efficient and proper operation of HVDC systems.

7.3.6 Protection and diagnosis of HVDC system faults

The modifications and enhanced complexity of power grid may alter the DC and AC side fault characteristics. Adaptive designing of protection schemes for HVDC systems effectively responds to the altering fault conditions and characteristics of faults. Extensive research is going on by considering various factors of power grid to effectively detect, classify and locate the faults in HVDC systems. The development and research efforts in HVDC technology, control design and practice in grid management are crucial to ensure the efficient and reliable integration of HVDC system into evolving power system.

7.4 Planning of HVDC Systems with Renewables

Renewable energy is a clean source available freely for harnessing in natural form. However, the issue of uncertain/intermittent output is one of the primary concerns of these resources. Their power output is dependent on natural events, not controllable and thus, cannot be dispatched. As such, the output power of wind and solar firms is decided by wind speed and solar isolation, respectively and both are uncertain in nature. Therefore, these uncertainties must be taken into account during planning for the mixed installation of renewable with thermal and hydro resources [27].

The addition of renewable energy resources must be supported by new transmission infrastructure to accommodate the uncertainty and gradually rise in demand. Thus, interconnecting the distributed resource to the transmission without discrimination should be considered by the system planners.

The independent system planner (ISP) cannot be biased from one type of generation technology to another.

On the contrary, ISP plans to provide reliable, low congestion costs and upgraded transmission facilities to avoid any violation of the reliability criterion. Economy and reliability are the key aspects of the planning of transmission services and interconnection of commercial generators.

The availability of renewable energy in typical locations and its dispersed and intermittent nature cause several issues with synchronous AC transmission systems. Such issues are power system stability, transmission economy, reliability, capacity, environmental concerns, congestion problems, etc. The integration and transmission projects of nonconventional energy also face challenges of the weak local grid, lack of reactive power support, and insufficient capacity. However, the HVDC transmission is an alternative and capable of overcoming these limitations. The unremitting improvement of power electronics and intellectual control systems increased the DC transmission voltage and reduced the cost and losses gradually. Thus, it is possible to transmit and integrate the large-scale renewable energies from the deep-water sea and onshore areas available where it's available in abundance.

The VSC-based HVDC technology is more likely to be used in future large-scale renewable energy projects [28]. China is leading in the world in integrating large-scale renewables using VSC and multi-terminal HVDC systems. The Zhangbei project in China is the world's first ±500 kV, four-terminal VSC-HVDC grid demonstrating the delivery of 4500 MW of clean energy. The United Kingdom and Germany are the next in commissioning offshore wind farms via HVDC interconnections in the North Sea [29]. To achieve the target of minimizing CO_2 emissions by 80% by 2040, the European Union is planning to establish a pan-European HVDC super-grid to deliver clean energy from offshore wind farms [30]. The idea is to encourage the growth of large-scale renewable energy from distant areas of Europe, ensure efficient access to clean energy, and restrain power interruption and fluctuations. This project also activates cross-border energy transactions and ensures the steady and secure operation of power grids. Such projects improve the coordination in the utilization and expansion of renewable energy and power systems, worldwide. However, the use of HVDC to transmit and integrate renewable sources also encountered several technical challenges such as:

7.4.1 Capacity optimization to reduce curtailment of RES and underutilization of HVDC system

The capacity of renewable energy integrated into the grid power via HVDC depends on either VSC or LCC converter capacity, mode of operation, or installed capacity of renewable plants. The incorrect ratio of the converter to plantinstalled capacity may lead to unnecessary shortening of renewable energy or underutilization of converter stations. Therefore, this ratio should be optimized to find out the utmost capacity of grid for integrating renewable energy. The HVDC projects of Shanghai Nanhui in China and Tjæreborg in Denmark were designed to realize such optimization.

7.4.2 Virtual generator or load shedding options to minimize frequency instability

The extensive addition of renewable energy sources (RES) leads to a decrease in power reserves during emergencies and inertia of a synchronized grid. The renewable sources integrated into the grid may barely respond to frequency change due to isolation of the HVDC system. This may lead to frequency instability of the power system. Thus, the appropriate options i.e., virtual generators and load shedding must be opted for to respond to the frequency instability.

7.4.3 Fault ride-through during islanding via HVDC

The DC grid will experience a power surplus; hence, a rise in voltage will result after the islanding operation of extensive renewable energy by means of VSC-HVDC due to line faults. This becomes more prominent with the uncontrolled generation of renewable energy especially when it is operating at rated capacity. Thus, the larger disturbance will be experienced at the receiving end of the AC network and result in blocking of the VSC converter at the same terminal due to commutation failure. Consequently, tripping out of entire renewable energy, due to quick rise in voltage and blocking of DC grid must be considered. Therefore, control coordination of HVDC with renewable plants for fault ride-through is a must.

7.4.4 Stability and safety threats due to sub-synchronous oscillation

The various components of the VSC-HVDC system generate harmonic frequency components, which may create resonance, and lead to sub-synchronous oscillation in systems integrated using the VSC-HVDC system. Thus, the stability and safety of the system are affected which may lead to a system outage. Therefore, oscillations of sub- or supersynchronous frequencies must be considered during the operation and planning of VSC-HVDC-based RES integration and transmission. An incident of harmonic oscillations observed in the German BorWin1 offshore wind power system in 2013, lead to a system outage [31].

7.5 Solutions using Soft Computing Techniques (SCT)

Soft computing techniques are the group of methodologies that aim to exploit the uncertainty and tolerance for imperfection to achieve robust, tractable, and low-cost solutions. Its main constituents are fuzzy logic (FL), artificial neural networks (ANNs), and genetic and evolutionary (GAE) algorithms [32]. This section covers the application of these soft computing techniques in the operation, control, and protection of an HVDC system.

In [33], authors investigate the performance of an online ANN-based current controller with choice to select different activation functions for the HVDC system. Basically, a small and large disturbance in reference current (I_{ref}), DC link current (I_{dc}), current error ($I_e = I_{ref} - I_{dc}$) are used for learning. The results show many attractive features of proposed inverse ANN-controller compared to a conventional PI-controller with respect to adaptability, damping, noise immunity, and ruggedness. However, the HVDC model used to test the performance is simple, linear, and experimental validation of the proposed ANN architecture is yet to be done. In [34], authors describe an intelligent ANN-FL current controller using heuristic and FL-based methods to adopt controller learning parameters. A comparative analysis showed that ANN-FL controller performed better when compared to traditional PI and ANN controller for fast and flexible control of HVDC link.

Traditional NN approaches for fault diagnosis in HVDC require huge training data sets for successful operation. In [35, 36], authors proposed an RMS pre-processor (RPP)-based NN to minimize the requirement of raw data sets for training. It uses a post-processor unit before fault classification (FC) to cleanup the noise of other spurious output signal present as shown in Figure 7.8. The block diagram in Figure 7.8 is applied for a simple 242 kV, 2 kA 1-pole HVDC systems, to detect and classify the symmetrical and unsymmetrical faults either on AC or DC sides.

The RPP uses three strategies to reduce raw data and improve the performance of NN-FC to obtain similar results as the classical NN approach. The

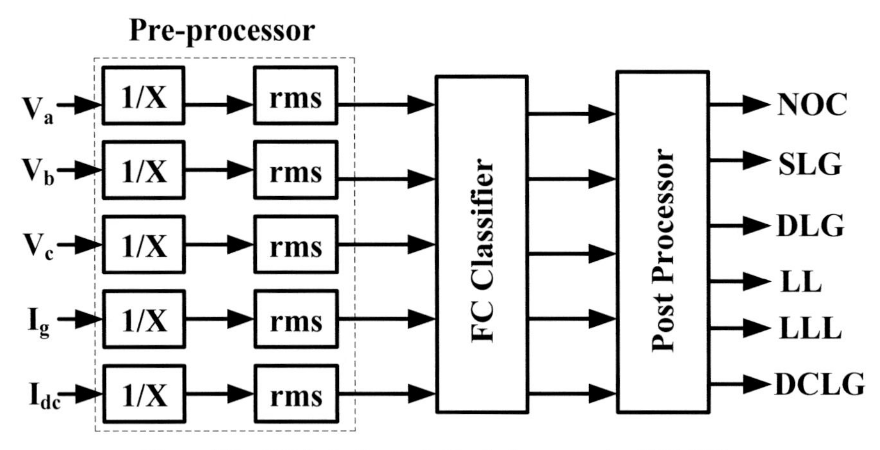

Figure 7.8 NN fault diagnosis block diagram using RPP [35].

first one is to use neutral current (i.e., sum of the current in three lines) as an alternative of individual currents of each line and line current of HVDC system instead of using three phase currents. The second strategy is to use an appropriate length window for covering pre- or postfault conditions to generate patterns for training. The third strategy is to generate synthetic training data set using expert knowledge of different fault scenarios such as double line-to-ground (DLG) and line-to-line (LL) faults.

The results show that the proposed method of training NN gives similar results as a classical NN trained with conventional full data set.

In [37], the concert of an HVDC system is improved during vibrant conditions by using neuro-fuzzy-based voltage-dependent current-order limiter (VDCOL). An HVDC system interfacing with a weak AC network encounters the problem of commutation failure. Under such a situation, an adaptive current order optimizes the system revival after fault and minimizes the events of commutation failure. The radial basis function (RBF) of NN output with fuzzy inference mechanism is used to adaptively select the reference DC current. The block diagram of an adaptive VDCL using neuro-fuzzy algorithm is tested for CIGRE benchmark model of an HVDC transmission system, as shown in Figure 7.9.

The dynamic conditions of starting of HVDC link, sag in DC link voltage, and improvement from fault and tracking reference current are considered over CIGRE benchmark model to check the concert of neuro-fuzzy-based VDCL (NF-VDCL) control unit. The comparison based I_{dc}, firing angle (α), I_{ref} and DC voltage (V_{dc}) for the conventional and NF-VDCL control

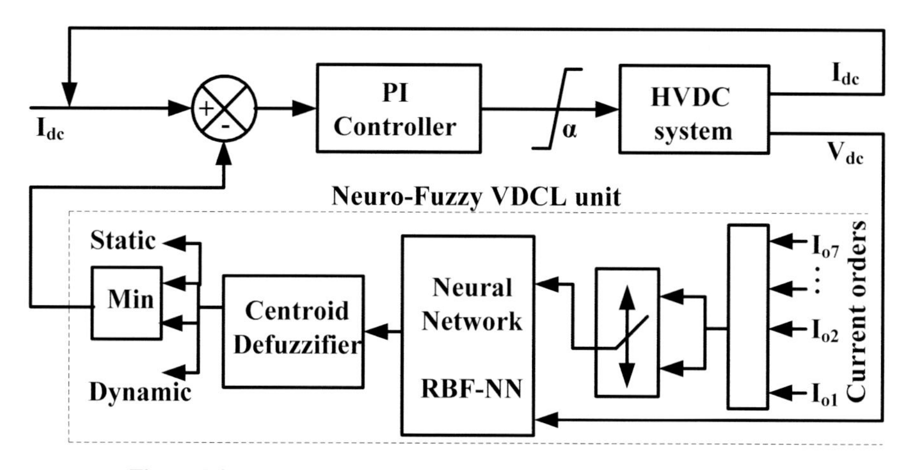

Figure 7.9 Neuro-fuzzy VDCL control unit for HVDC system [37].

is shown in Figure 7.10. Results show that start-up is fast in conventional VDCL and hits α-minimum limit to 9°. The regulator quickly recovered from α-minimum to final value of 16° within 0.125 s. Recovery of DC current is smooth with 90% settling time of 0.2 s. However, the NF-VDCL-based HVDC control achieves better start-up with availability of α-minimum limit and no spikes in I_{ref}. The I_{dc} value reaches to 90% level smoothly, like the conventional VDCL.

A three-phase short circuit at the AC side of the inverter terminal is used for simulating a voltage dip to zero. Both NF and conventional VDCL control can minimize I_{ref} within 20–40 ms of their limited values. The result is shown in Figure 7.11, illustrating that the response for NF is slightly faster with a higher minimum range of I_{ref} up to 0.1 p.u.

A comparative analysis of a conventional and NF-VDCL during fault recovery is indicated in results of Figure 7.12. It indicates that the conventional VDCL unit has larger oscillations in DC line current, reference current, DC link voltage, or in firing angle as compared to the NF-VDCL unit. The higher oscillations in I_{ref} close to 0.8 p.u. cause the risk of commutation failure in HVDC system. However, the NF exhibits no such oscillation.

HVDC is known for fast controllability of desired power flow or modulating DC power to improve stability of adjacent AC systems. The verification of fast controllability is measured using a reference current tracking performance of VDCL controller. Thus, a 20% drop in I_{ref} over a period of 28 cycles initiated from 0.8 s is simulated with NF-VDCL and without VDCL

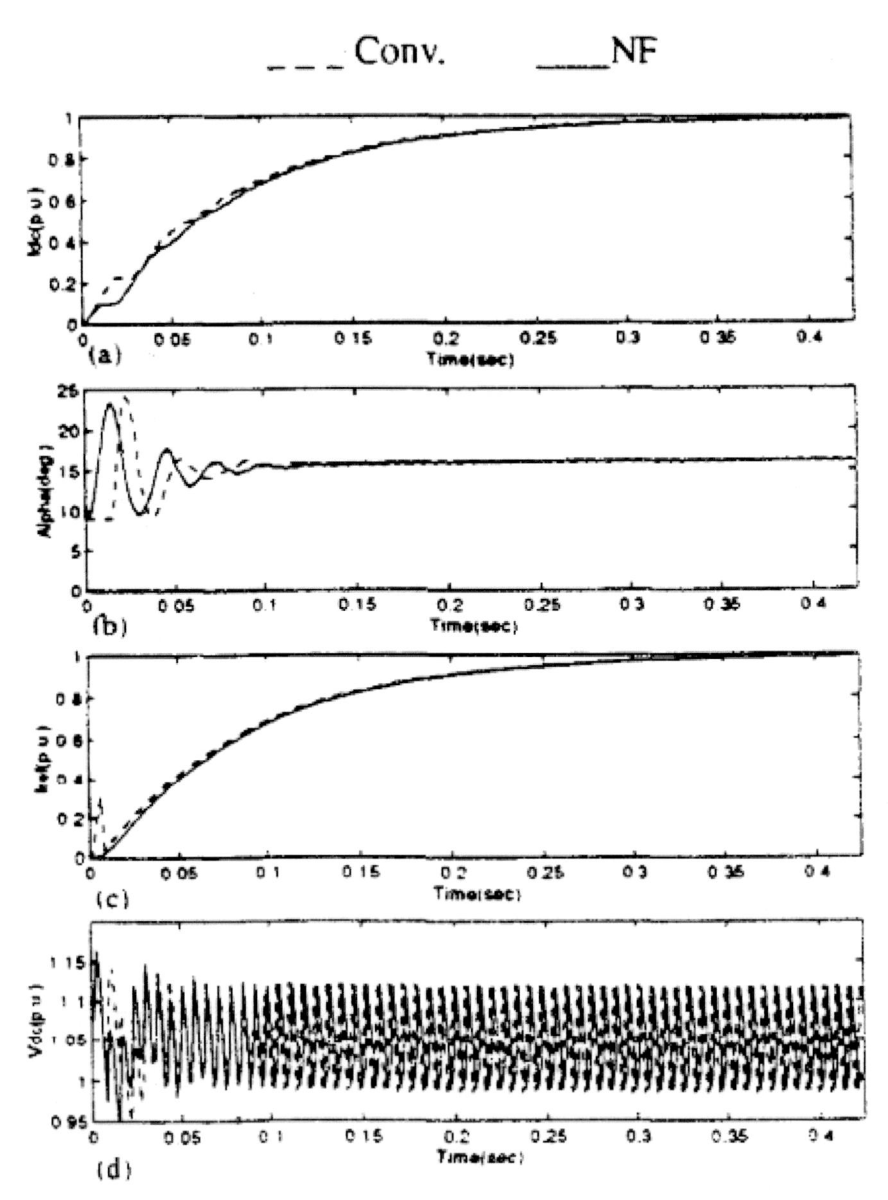

Figure 7.10 Dynamic response of DC link during staring-up of HVDC system [37].

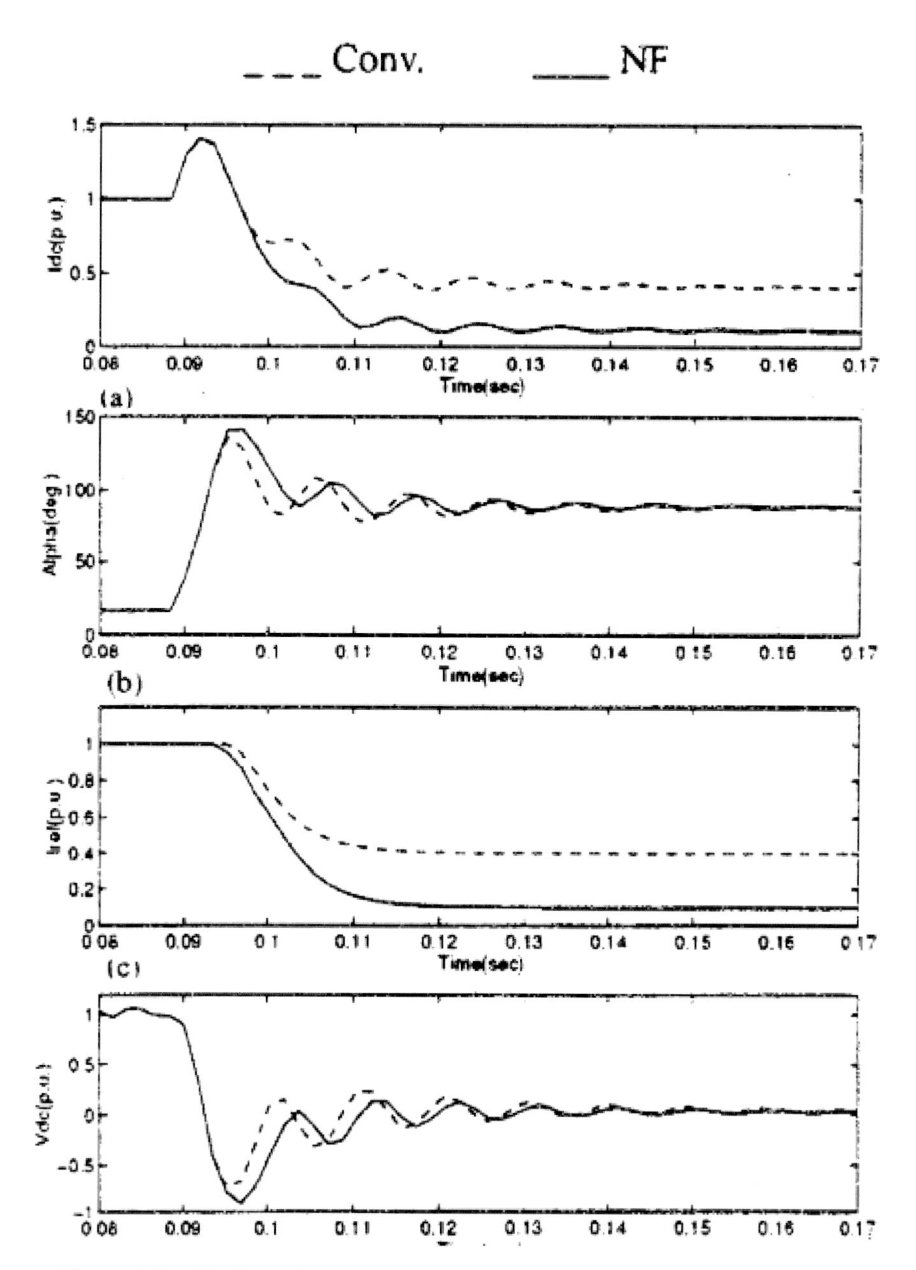

Figure 7.11 VDCL performance during three-phase fault at inverter terminals [37].

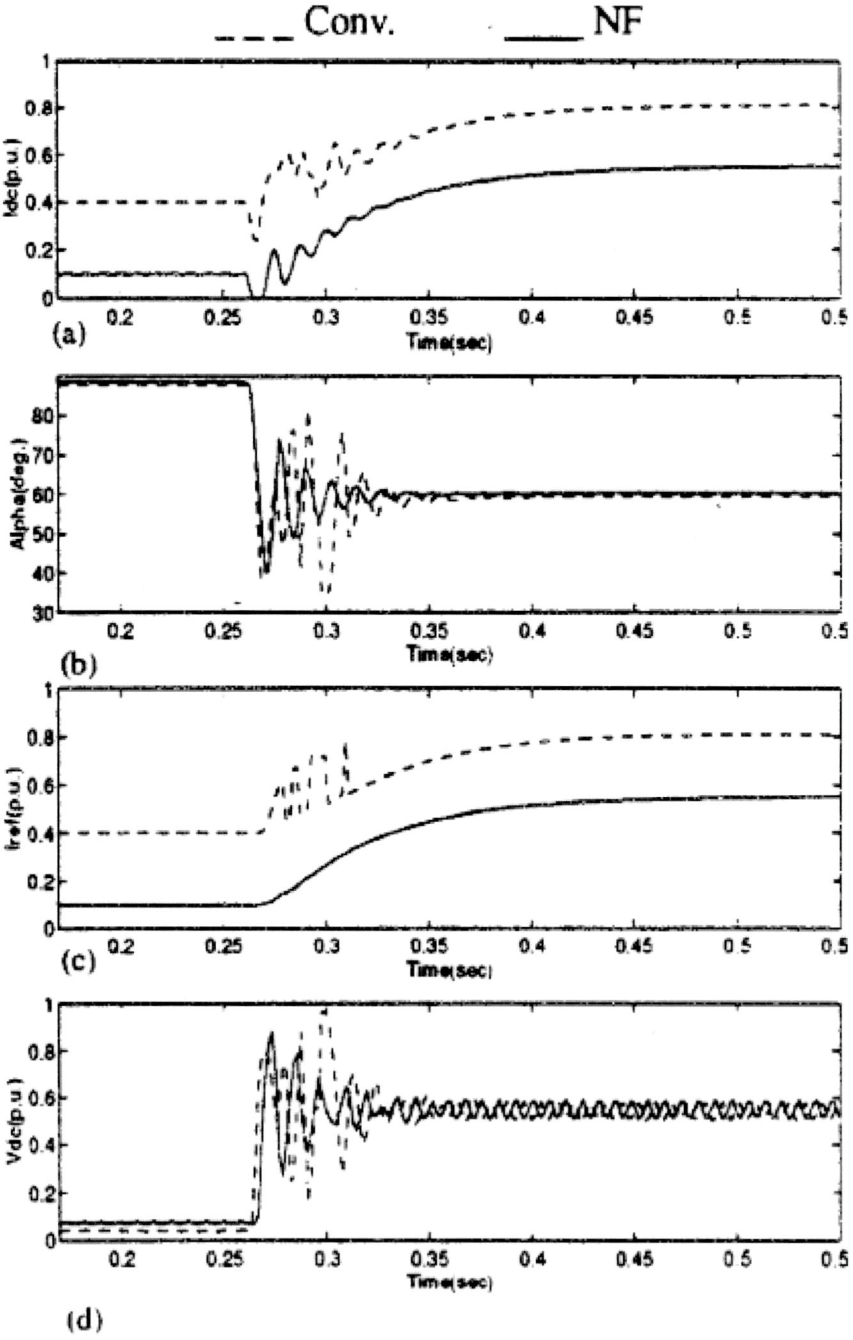

Figure 7.12 Performance comparison of conventional and NF-VDCL units during fault recovery [37].

unit. The resultant DC link current, I_{ref}, firing angle and V_{dc} are shown in Figure 7.13. It indicates that NF-VDCL control can damp out oscillations during tracking with no overshoot.

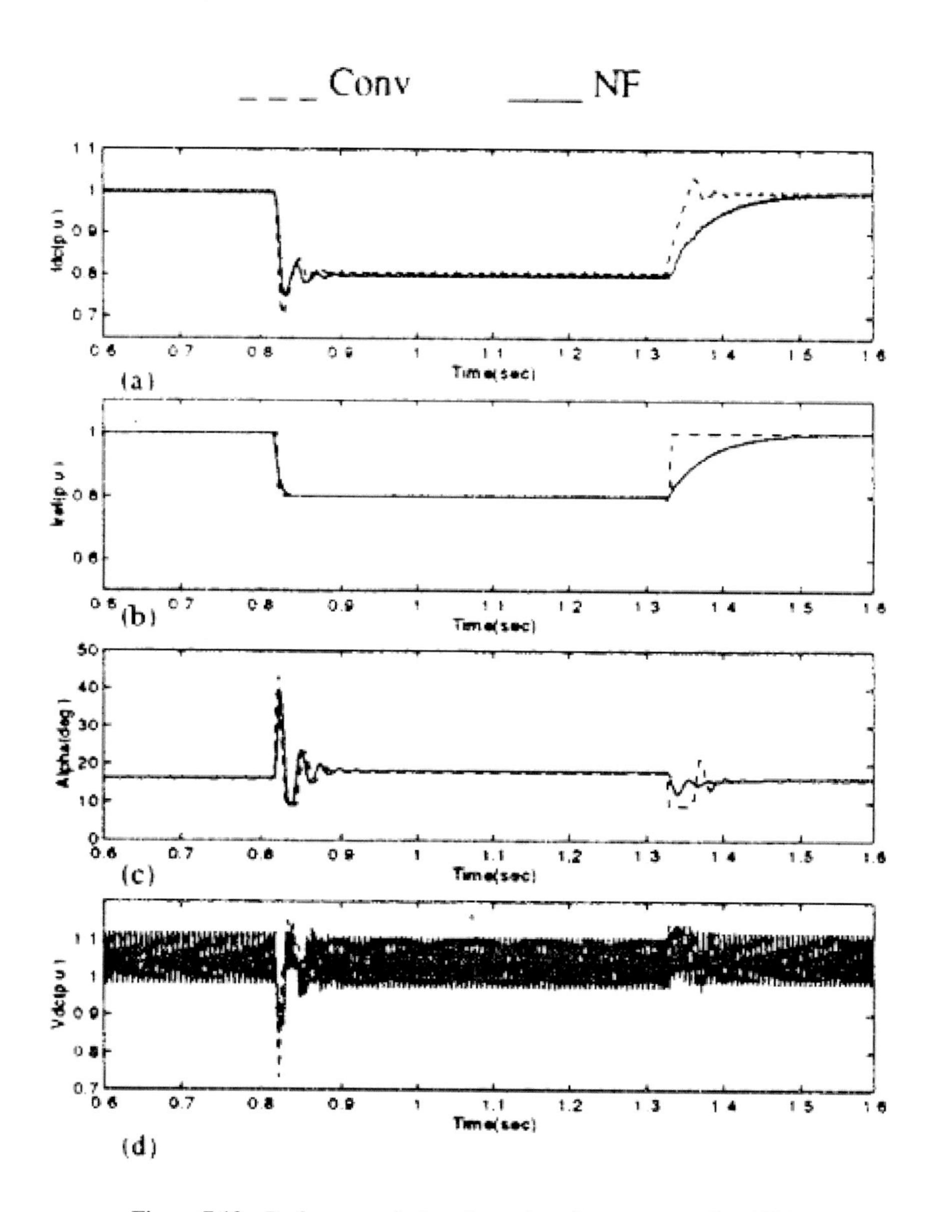

Figure 7.13 Performance during change in reference power flow [37].

In [38], the authors proposed a three counter propagation network (CPN)-based NN architecture to detect six types of AC-DC faults in HVDC system. Based on reliable fault detection, an optimized control can be suggested to improve the dynamic performance. The three CPN-NN models are developed based on the combination of inputs such as AC and DC node voltages with or without phase angle, RMS or instantaneous, or rectifier neutral bus three-phase voltages. The input used for three architectures are: (1) RMS value of each phase voltage, (2) RMS phase voltages together with phase angles, (3) sampled instantaneous phase voltages. The model based on the third type of input selection reduces the response time from $1-2$ cycles to a half-cycle. However, training time and complexity increase due to large neurons on the Kohonen layer. It can differentiate double line (DL) faults from double line-to-ground (DLG) at the remote-end which is distinguished in first or second type of input data. So, if such distinction is not required then method 1 is the most simple and practical to implement and the delay of $1-2$ cycles is acceptable in most HVDC systems.

In [39], authors present a signal decomposition approach based on wavelet multi-resolution analysis (MRA) to extract features for monitoring disturbances in HVDC systems. Both AC and DC side signals are used to extract the promising features capable of classifying various disturbances in the HVDC system. The feat of the projected technique is examined for the following four cases: (1) Normal operating condition (NOC), (2) DC line faults (DCLG), (3) commutation failure (CF), and (4) one line-to-ground (OLG) faults, using CIGRE HVDC benchmark model. The signal used for MRA is the rectifier- and inverter-side phase voltages, V_{dc} and I_{dc}, as shown in Figure 7.14.

The signal's sampling frequency is 4 kHz decomposed over a window size of 8 cycles into 9 levels. Each level consists of a particular frequency band as shown in Table 7.2. The different disturbance types have a particular pattern irrespective of the duration of disturbances.

Table 7.2 Signal resolution and respective frequency band [39].

Decomposition levels	**Frequency band**	**Decomposition levels**	**Frequency band**
1	1–2 kHz	6	31.25–62.5 kHz
2	0.5–1 kHz	7	15.75–31.25 Hz
3	0.250–0.5 kHz	8	7.875–15.75 Hz
4	0.125–0.250 kHz	9	3.937–7.875 Hz
5	0.0625–0.125 kHz		

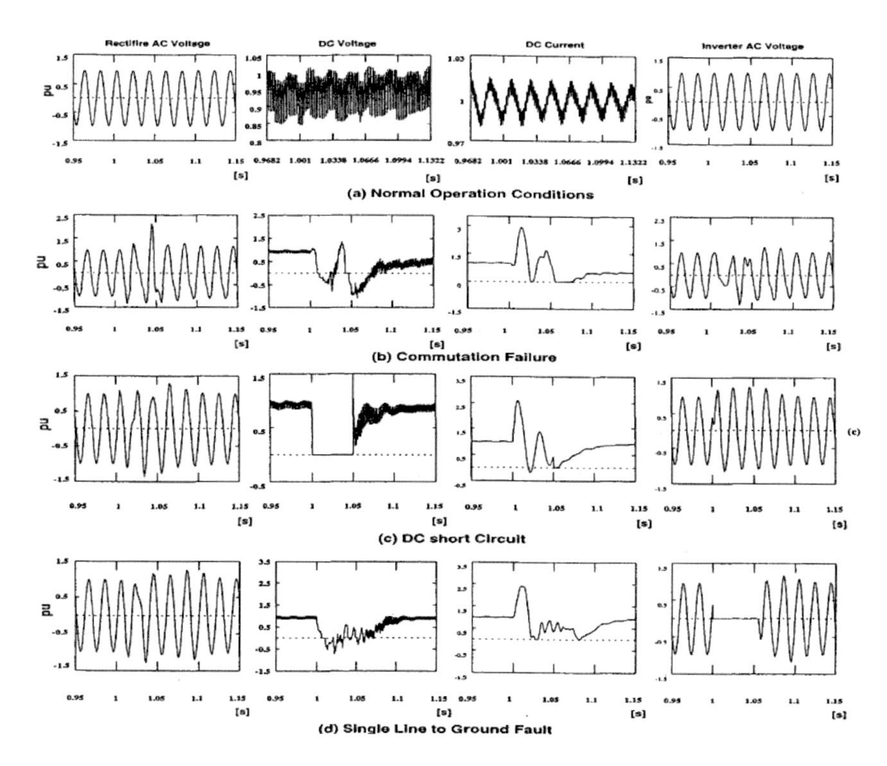

Figure 7.14 HVDC system parameters for MRA during (a) NOC, (b) CF, (c) DSLG, and (d) SLG fault [39].

The rules to classify the disturbance are constituted by evaluating energy deviation of detailed feature vectors obtained from wavelet MRA. The mathematical expression for energy deviation (ΔE) between fault and NOC is given by eqn (7.7).

$$\Delta E_{(j)} = ||d_{(j)}||_{disturbance\,type} - ||d_{(j)}||_{normal\,operation\,(NOC)}, \tag{7.7}$$

where *d*(*j*) is the detailed wavelet coefficient of *j*th resolution level, used to generate translational invariant features. The developed rule classifying disturbances are given as follows in Table 7.3.

Table 7.3 MRA-based rules for classifying disturbance.

If	$\Delta E_{(6)} = 0$	&	$\Delta E_{(8)} = 0$	$\leftrightarrow$	NOC
If	$\Delta E_{(6)} > 0$	&	$\Delta E_{(8)} > 0$	$\leftrightarrow$	SLG
If	$\Delta E_{(6)} < 0$	&	$\Delta E_{(8)} > 0$	$\leftrightarrow$	CF
If	$\Delta E_{(6)} < 0$	&	$\Delta E_{(8)} = 0$	$\leftrightarrow$	DCLG

7.6 Real-time (RT) implementation and hardware-in-the-loop (HIL) simulations

A summary of the discoveries and identification of the advances or contributions in HVDC systems is provided by this work. In the field of integration of HVDC systems, there exist multi-domain behaviour that creates challenges in smart grid operation, control, and protection designing. In terms of simulation, modelling, and testing, their validation and implementation necessitate careful consideration. The conventional strategy of concentrating on a single object, real hardware, or a realistic model while considerably reducing the rest of the system under test is no longer adequate. Real-time implementation and hardware-in-the-loop (HIL) simulation have emerged as essential tools for verifying the behaviour of HVDC systems and their interaction/impact with the rest of the power system. This section overviews the status of the RT-HIL approach for HVDC implementation in smart grid to perform the experiments and model validation. The current limitations and necessary future developments in HIL are also highlighted here.

The point-to-point or multi-terminal HVDC network is growing due to bulk power transmission and the integration of solar and wind farms into the conventional power system. This has created challenges in the operation, stability, control, and protection of the modern power system. This leads to the need for validation and testing of innovative solutions or devices integrated into the traditional power system. Real-time simulation, particularly the usage of hardware-in-the-loop, is a focus of the task force on "Real Time Simulation of Power and Energy Systems" of IEEE Power and Energy Society. It is found that HIL techniques can meet a variety of requirements related to the assessment of power system stability, fast prototyping, and proper protection designing to assure the security and safety of power systems with novel solutions. The RT-HIL approach combines an actual hardware setup with a simulation tool to permit testing of software or hardware components beneath realistic environments. The simulator's carrying out requires extremely small time-steps to adhere to the RT constraints of the physical system. However, the HIL allows the replacement of faulty or incomplete, or inaccurate models with their counterparts present in the real-world [40]. The HIL techniques have been effectively employed in smart grids for validating various experiments, devices such as converter topologies [41], protective relays [42], and innovative devices for the entire energy systems [43].

Though HIL testing offers extensive potential for validating innovative solutions in smart grids that integrate distributed energy sources. The devices

with multiple behaviours, the HIL is still being developed and suffers from some limitations such as complexity in the integration of communication layers with HIL, limited simulation capacity for complex systems, distributed and remote HIL for combined experiments, etc. Furthermore, a lack of standardization in the HIL approach leads to the use of proprietary interfaces that restrict a generic framework to promote the reusability of trials or experiments.

7.6.1 Hardware-in-the-loop techniques for smart grids

The controller hardware-in-loop (CHIL) and power hardware-in-loop (PHIL) are the two basic classifications of the HIL approach in a smart grid. CHIL technique includes the testing of controllers used for power converters, where the hardware under test (HUT) and real-time simulator exchange signals via its information (digital-to-analogue or analogue-to-digital) ports. The phasor measurement units (PMUs), relays, or monitoring devices coupled to the RT-simulators are also classified as CHIL, beside controllers. Such protection and control equipment are verified in closed loop underneath different fault and dynamic conditions to enhance their acceptability for energy and power components. PHIL, on the other hand, includes the testing of devices that absorb or generate power such as photovoltaic inverters. As a result, a power interface is essential for PHIL. Thus, the power interface in PHIL, RT-simulator, and HUT are the three main elements of the HIL approach for validating novel integrated devices in a smart grid.

The RT-simulator provides real-time computation and I/O capability to emulate behaviours of the simulated models beneath transient conditions. The simulator offers considerable flexibility in the design and execution of various testing scenarios. Most RT-simulators use multi-processors operating in parallel to develop a platform for running simulations in real-time and I/O ports are used to interface with peripheral hardware. The Simulink model is developed and compiled off-line on the host computer and loaded to the targeted platform and the real-time simulation results are monitored on a digital storage oscilloscope (DSO) or host computer as well.

A modern VSC-HVDC system used for integrating asynchronous AC grids or sources of renewable energies i.e., wind farms or PV system to the existing AC grid is shown in Figure 7.15. The detailed system description and equipment capacity are mentioned in Appendix A in Table A1 and Table A2 respectively. Initially, the off-line modelling of a 200 MW, $\pm$ 100 kV bipolar point-to-point VSC-HVDC system, integrating an AC system of

50 Hz frequency to another AC system of 60 Hz frequency, was done. The voltage and power ratings of both systems are 230 kV, 2000 MVA, developed on the MATLAB/Simulink environment on the host computer. Afterward, the developed Simulink VSC-HVDC model was deployed on the OPAL-RT (OP 4510) real-time simulator for validation and real-time simulation of VSC-HVDC fault characteristics. Three different transient scenarios with external and internal fault conditions of the system shown in Figure 7.15 are simulated in real-time on the OPAL-RT platform. The real-time simulation results are obtained on DSO and can be used to design protection or control algorithms and validate the developed MATLAB/Simulink model.

Figures 7.16 and 7.17 are RT-simulation results of measured AC-DC line parameters concerning external fault F_1 and F_3, respectively. Figure 7.18 shows OPAL-RT results for internal fault F_2. Figure 7.19 represents the complete OPAL-RT simulation platform including the host computer and DSO interfacing via input-output (I/O) ports of the RT-simulator. The DSO used is a 4-channel output device recording reactive power (Q_{PCC-1}) at bus B_1, active power (P_{dc}), positive pole-to-ground (V_{Pdc}), and negative pole-to-ground voltage (V_{Ndc}) on channels 1, 2, 3, and 4, respectively. The analogue or digital I/O ports of OPAL-RT interface with DSO via ethernet cable. The test system of Figure 7.15 is modelled with a fixed-step, ode-4 (Runge–Kutta) solver in MATLAB/Simulink software. A sampling frequency of 50 kHz or lower is preferred because of the real-time simulation constraints of OPAL-RT.

Figures 7.16 and 7.17 show the dynamics in Q_{PCC-1}, P_{dc}, $V_{Pdc,}$ and V_{Ndc} at channels 1, 2, 3, and 4 of DSO during three-phase external faults F_1 on bus B_1 at station 1 and three-phase fault F_3 at bus B_2 of station 2, respectively. The fault exists for six cycles of the fundamental frequency of 50 Hz, indicating a drop in P_{dc} during external fault F_1 shown in Figure 7.16, which recovered within a fraction of the cycle after fault clearance. The

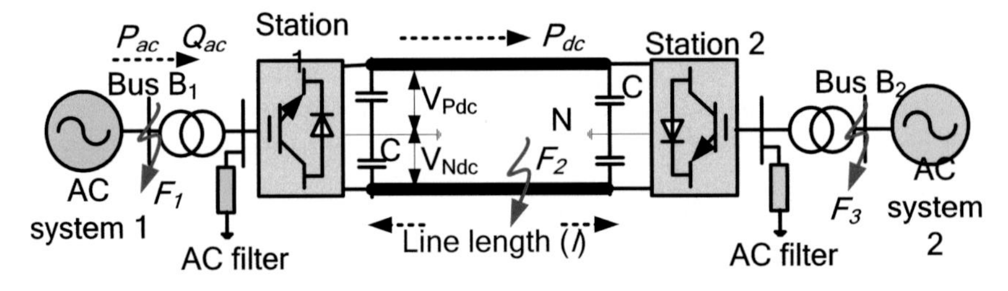

Figure 7.15 HVDC test system for hardware-in-loop simulation.

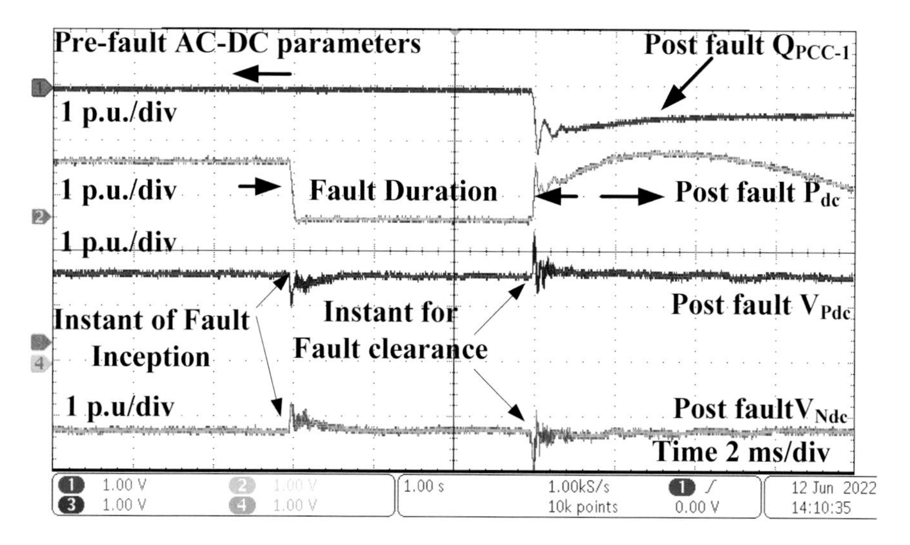

Figure 7.16 Real-time OPAL-RT simulator during external faults F_1.

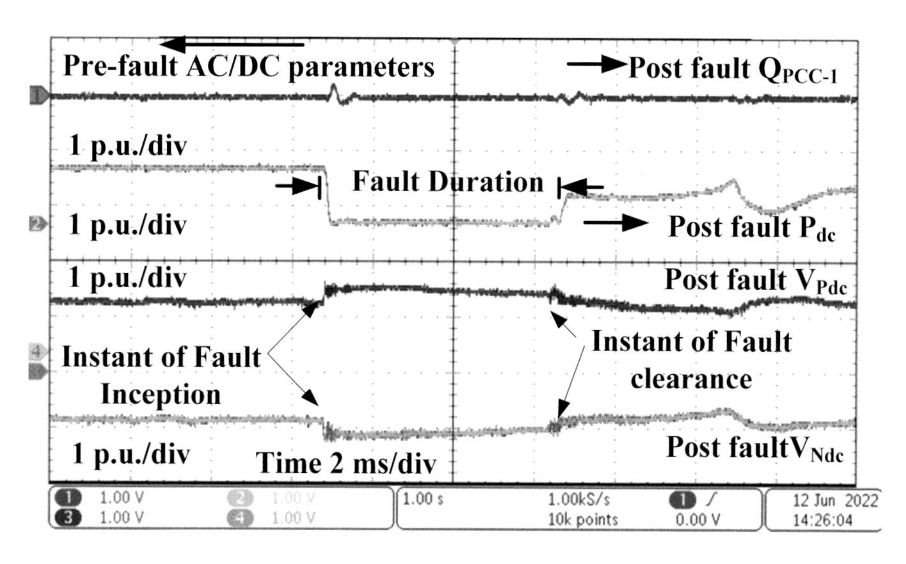

Figure 7.17 Real-time OPAL-RT simulator during external faults F_3.

40 Mvar filter bank installed at station 1 and independent real and reactive power of VSC need zero reactive power demand from AC system 1 as indicated in Q_{PCC-1} on pre-fault and during fault conditions. However,

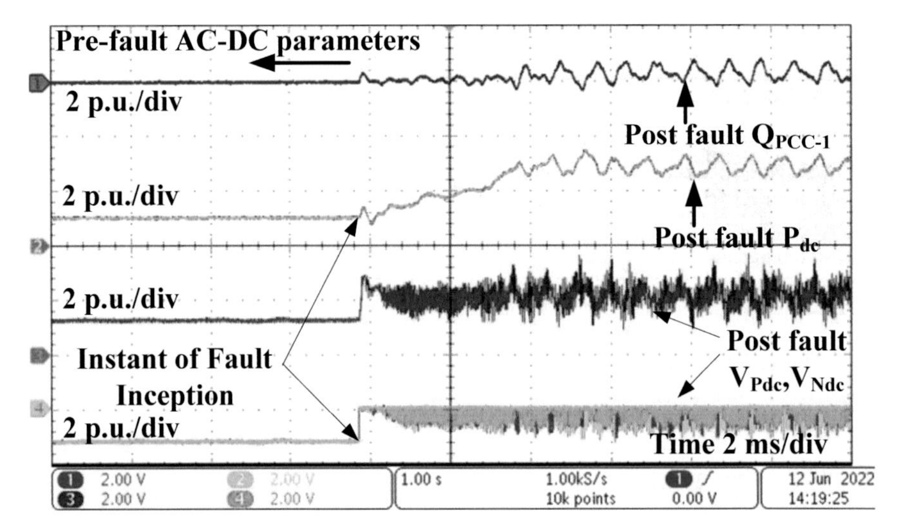

Figure 7.18 OPAL-RT result during internal pole-to-ground fault F_2.

the sharp transient demand in Q_{PCC-1} appears from AC system 1 at post-fault condition to establish the active power flow via the HVDC link. A minor dynamic in DC link voltage on each pole V_{Pdc} and V_{Ndc} appear at the instance of fault inception and clearance in the opposite direction of its absolute value. Meanwhile, the DC link voltage is maintained constant during fault. Similarly, the results in Figure 7.17 indicated negligible fluctuation in reactive power Q_{PCC-1} at B_1 while the P_{dc} completely halted during fault F_3. Due to sudden interruption in P_{dc} and overcharging of DC link capacitor, a small swell in V_{Pdc} and V_{Ndc} appears.

Furthermore, the results in Figure 7.18 show that an internal DC line fault F_2 (negative pole-to-ground) causes total power interruption from the faulty pole during fault and post-fault conditions. This is due to the operation of the faulty pole into capacitor discharging and the diode freewheeling stage of VSC after a fault. However, the P_{dc} for the healthy pole rises with larger fluctuation causing overloading and stress to the healthy pole. The DC link capacitors of the healthy pole experience overcharging hence, sustained transient overvoltage appears on the healthy pole (V_{Pdc}) of the HVDC link. However, a transient DC link voltage of a faulty pole (V_{Ndc}) exists with an average magnitude close to zero. The small fluctuation in reactive power demand from AC system 1 appears with fluctuation in P_{dc} and V_{Pdc}, as appears in channel 1 of Figure 7.18.

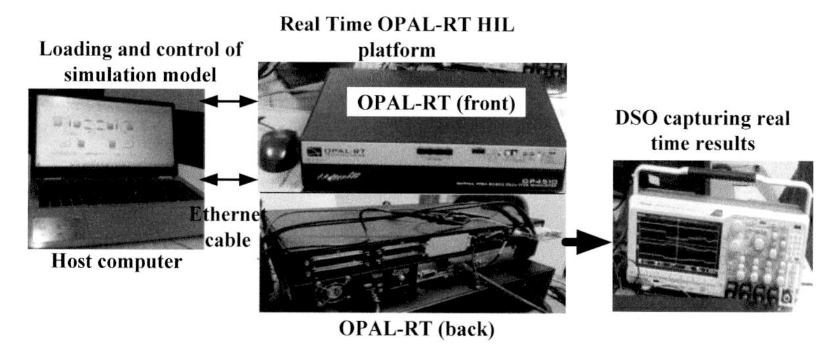

Figure 7.19 Real-time OPAL-RT simulation setup for HVDC system.

7.7 Conclusion

Traditionally, the power transmission technology is based on an HVAC transmission system which provides satisfactory operation with good levels of reliability and availability. However, there are challenges primarily arising from long-distance transmission and massive integration of renewable sources, which leads to vulnerability and unpredictability in grid operation. The HVDC systems are an alternate technique to overcome the limitation of the HVAC systems during point-to-point or multi-terminal power transmission. This chapter provides a summary of various issues existing with HVDC systems and the application of soft computing techniques to resolve such issues. The HVDC system, its classic control arrangements, and operational modes are discussed. The various issues related to transient and small-signal stabilities are also covered. The integration of renewable sources using VSC-HVDC is now widespread. The highlighted contemporary HVDC issues arise due to the ongoing evolution in the power system to serve societal needs. The fuzzy logic, ANN, and wavelet MRA are some of the soft computing techniques that are now emerging for the protection and control requirements of HVDC systems. Furthermore, to validate the various simulation models of power and control, PHIL and CHIL hardware-in-the-loop real-time simulator tools are required to analyse the fault characteristics with the real-time hardware-in-the-loop arrangements.

References

[1] Y. Li et al., "Fault Ride-through Demand of Large-scale Islanded Renewable Energy Connected to VSC-HVDC System and Its Key

Technologies," 2020 4th Int. Conf. HVDC, HVDC 2020, pp. 1248–1252, 2020, doi:10.1109/HVDC50696.2020.9292778.

[2] D. Obradovic, M. Oluic, R. Eriksson, and M. Ghandhari, "Supplementary Power Control of an HVDC System and Its Impact on Electromechanical Dynamics," IEEE Trans. Power Syst., vol. 36, no. 5, pp. 4599–4610, Sep. 2021, doi:10.1109/TPWRS.2021.3056763.

[3] R. Wachal, "Life extension issues for HVDC control systems," IET Conf. Publ., vol. 2010, no. 570 CP, 2010, doi:10.1049/cp.2010.0972.

[4] Y. Zhang, H. Ding, and R. Kuffel, "Key techniques in real time digital simulation for closed-loop testing of HVDC systems," CSEE J. Power Energy Syst., vol. 3, no. 2, pp. 125–130, 2017, doi:10.17775/cseejpes.2017.0016.

[5] M. H. Okba, M. H. Saied, M. Z. Mostafa, and T. M. Abdel-Moneim, "High voltage direct current transmission - A review, part I - Converter technologies," 2012 IEEE Energytech, Energytech 2012, vol. 2, 2012, doi:10.1109/EnergyTech.2012.6304651.

[6] O. E. Oni, K. I. Mbangula, and I. E. Davidson, "A Review of LCC-HVDC and VSC-HVDC Technologies and Applications," Trans. Environ. Electr. Eng., vol. 1, no. 3, p. 68, Sep. 2016, doi:10.22149/teee.v1i3.29.

[7] M. Schenk et al., "State of the Art of Bipolar Semiconductors for Very High Power Applications Key Design and Technology of Bipolar Semiconductors for Ultra High Power Applications," no. May, pp. 19–21, 2015.

[8] J. Vobecky, V. Botan, K. Stiegler, and U. Meier, "A Novel Ultra-Low Loss Four Inch Thyristor for UHVDC," 2015 IEEE 27th Int. Symp. Power Semicond. Devices IC's, pp. 413–416, doi:10.1109/ISPSD.2015.7123477.

[9] J. Vobecký, "THE CURRENT STATUS OF POWER SEMICONDUCTORS? Jan Vobecký," Facta Univ. Ser. Electron. Energ., vol. 28, no. June 2015, pp. 193–203, 2015, doi:10.2298/FUEE1502193V.

[10] J. He, M. Li, Y. Jun, C. Qing, X. Tao, and Y. Zhao, "Research on dynamic characteristics and countermeasures of AC-DC hybrid power system with large scale HVDC transmission," POWERCON 2014 - 2014 Int. Conf. Power Syst. Technol. Towar. Green, Effic. Smart Power Syst. Proc., no. Powercon, pp. 799–805, 2014, doi:10.1109/POWERCON.2014.6993874.

[11] Y. Jiang-Hafner, H. Duchen, M. Karlsson, L. Ronstrom, and B. Abrahamsson, "HVDC with voltage source converters - A powerful standby

black start facility," Transm. Distrib. Expo. Conf. 2008 IEEE PES Powering Towar. Futur. PIMS 2008, 2008, doi: 10.1109/TDC.2008.4517039.

[12] O. Kotb and V. K. Sood, "A Hybrid HVDC Transmission System Supplying a Passive Load," 2010 IEEE Electr. Power Energy Conf., no. May, pp. 1–6, doi:10.1109/EPEC.2010.5697183.

[13] C. Zhao and C. Guo, "Complete-independent control strategy of active and reactive power for VSC based HVDC system," 2009 IEEE Power Energy Soc. Gen. Meet. PES '09, pp. 1–6, 2009, doi:10.1109/PES.2009.5275743.

[14] Massoud Amin, A. M. Annaswamy, C. L. Demarco, and T. Samad, "IEEE Vision for Smart Grid Controls: 2030 and Beyond Reference Model," pp. 1–10, 2013.

[15] X. Jin, Y. Zhang, and X. Wang, "Strategy and coordinated development of strong and smart grid," 2012 IEEE Innov. Smart Grid Technol. - Asia, ISGT Asia 2012, no. 1, pp. 12–15, 2012, doi:10.1109/ISGT-Asia.2012.6303208.

[16] I. Colak, R. Bayindir, and S. Sagiroglu, "The Effects of the Smart Grid System on the National Grids," 8th Int. Conf. Smart Grid, icSmartGrid 2020, pp. 122–126, 2020, doi:10.1109/icSmartGrid49881.2020.9144891.

[17] M. Barnes, D. Van Hertem, S. P. Teeuwsen, and M. Callavik, "HVDC Systems in Smart Grids," Proc. IEEE, vol. 105, no. 11, pp. 2082–2098, 2017, doi:10.1109/JPROC.2017.2672879.

[18] T. Bandaru, U. Dhawa, D. Chatterjee, and T. Bhattacharya, "Improving the Transient Stability by Modifying the Power Exchange by the HVDC Transmission," 2018 20th Natl. Power Syst. Conf. NPSC, no. 2, doi: 10.1109/NPSC.2018.8771725.

[19] Z. Hu, C. Mao, and J. Lu, "Improvement of transient stability in AC system by HVDC Light," Proc. IEEE Power Eng. Soc. Transm. Distrib. Conf., vol. 2005, no. 4, pp. 1–5, 2005, doi:10.1109/TDC.2005.1546821.

[20] A. Nakamura et al., "Stabilization of Power System including Wind Farm by Cooperative Operation Between HVDC Interconnection Line and Battery," 2020 IEEE Int. Conf. Power Energy, pp. 60–65, 2022, doi:10.1109/PECon48942.2020.9314604.

[21] S. Barsali, M. Ceraolo, P. Pelacchi, and D. Poli, "Use of VSC-HVDC links for power system restoration," 2019 AEIT HVDC Int. Conf. AEIT HVDC 2019, 2019, doi:10.1109/AEIT-HVDC.2019.8740521.

[22] P. Thepparat, D. Retzmann, E. Ogee, and M. Wiesinger, "Smart transmission system by HVDC and FACTS," 2013 IEEE Grenoble Conf.

PowerTech, POWERTECH 2013, 2013, doi:10.1109/PTC.2013.665 2137.

[23] A. M. and A. B. Feng Wang, Lina Bertling, Tuan Le, "An Overview Introduction of VSC-HVDC: State-of-art and Potential Applications in Electric Power Systems," vol. httpÂă: //w, no. 21, rue d'Artois, F-75008 PARIS, 2011, [Online]. Available: https://core.ac.uk/download/pdf/70 601345.pdf.

[24] R. K. Antar, B. M. Saied, G. A. Putrus, and R. A. Khalil, "Treating the Impacts of Connecting HVDC Link Converters with AC Power System Using Real-Time Active Power Quality Unit," e-Prime, vol. 1, no. September, p. 100013, 2021, doi:10.1016/j.prime.2021.100013.

[25] G. Pinares, "Analysis of the DC Dynamics of VSC-HVDC Systems Connected to Weak AC Grids Using a Frequency Domain Approach," 2014 Power Syst. Comput. Conf., doi:10.1109/PSCC.2014.7038406.

[26] H. Zhang and A. Background, "A New Strategy of HVDC Operation for Maximizing Renewable Energy Accommodation," 2017 IEEE Power Energy Soc. Gen. Meet., doi:10.1109/PESGM.2017.8274477.

[27] J. Lin, "Integrating the First HVDC-Based Offshore Wind Power into PJM System-A Real Project Case Study," IEEE Trans. Ind. Appl., vol. 52, no. 3, pp. 1970–1978, 2016, doi:10.1109/TIA.2015.2511163.

[28] Y. Li, H. Liu, X. Fan, and X. Tian, "Engineering practices for the integration of large-scale renewable energy VSC-HVDC systems," Glob. Energy Interconnect., vol. 3, no. 2, pp. 149–157, 2020, doi:10.1016/j.gl oei.2020.05.007.

[29] S. W. Ali et al., "Offshore Wind Farm-Grid Integration: A Review on Infrastructure , Challenges , and Grid Solutions," IEEE POWER ENERGY Soc. Sect., vol. VOLUME 9, pp. 102811–102827, 2021, doi:10.1109/ACCESS.2021.3098705.

[30] A. L. Abbate et al., "Evolution of the pan-European transmission system towards the development of a prospective global power grid," 2019 16th Int. Conf. Eur. Energy Mark., doi:10.1109/EEM.2019.8916271.

[31] C. Buchhagen, C. Rauscher, A. Menze, and J. Jung, "BorWin1 - First experiences with harmonic interactions in converter dominated grids," Int. ETG Congr. 2015; Die Energiewende - Blueprints New Energy Age, pp. 27–33, 2015.

[32] U. Kumari, "Soft computing applications: A perspective view," Proc. 2nd Int. Conf. Commun. Electron. Syst. ICCES 2017, vol. 2018-Janua, no. Icces, pp. 787–789, 2018, doi:10.1109/CESYS.2017.8321190.

[33] K. G. Narendra, V. K. Sood, K. Khorasani, and R. V. Patel, "Investigation into an artificial neural network based on-line current controller for an HVDC transmission link," IEEE Trans. Power Syst., vol. 12, no. 4, pp. 1425–1431, 1997, doi:10.1109/59.627837.

[34] K. G. Narendra, V. K. Sood, K. Khorasani, and R. V. Patel, "Intelligent current controller for an HVDC transmission link." IEEE, Columbus, OH, USA, 1997, doi:10.1109/PICA.1997.599379.

[35] H. Etemadi, V. K. Sood, K. Khorasani, and R. V. Patel, "Neural network based fault diagnosis in an HVDC system," Int. Conf. Electr. Util. Deregul. Restruct. Power Technol. 2000, City Univ. London, 4-7 April 2000., no. Idc, pp. 209–214, 2002, doi:10.1109/drpt.2000.855665.

[36] K. G. Narendra, V. K. Sood, R. V. Patel and K. Khorasani, "Application of a radial basis function (RBF) neural network for fault diagnosis in a HVDC system," IEEE Trans. Power Syst., vol. 13, no. 1, pp. 177–183, 1998, doi:10.1109/59.651633.

[37] K. G. Narendra and K. Khorasani, "A Neuro-Fuzzy VDCL Unit to Enhance the Performance of an HVDC System," Proc. 1995 Can. Conf. Electr. Comput. Eng., p. 441, 1995, doi:10.1109/CCECE.1995.528169.

[38] N. Kandil, K. Khorasani, R. V. Patel, and V. K. Sood, "Fault identification in an AC-DC transmission system using neural networks," IEEE Trans. Power Syst., vol. 7, no. 2, pp. 812–819, 1992, doi:10.1109/59.141790.

[39] A. M. Gaouda, E. F. El-Saadany, M. M. A. Salama, V. K. Sood, and A. Y. Chikhani, "Monitoring HVDC systems using wavelet multi-resolution analysis," IEEE Trans. Power Syst., vol. 16, no. 4, pp. 662–670, 2001, doi:10.1109/59.962411.

[40] V. H. Nguyen et al., "Real-Time Simulation and Hardware-in-the-Loop Approaches for Integrating Renewable Energy Sources into Smart Grids: Challenges & Actions," 2017, [Online]. Available: http://arxiv.org/abs/1710.02306.

[41] P. Kotsampopoulos, V. Kleftakis, G. Messinis, and N. Hatziargyriou, "Design, development and operation of a PHIL environment for Distributed Energy Resources," IECON Proc. (Industrial Electron. Conf., pp. 4765–4770, 2012, doi:10.1109/IECON.2012.6389005.

[42] M. S. Almas, R. Leelaruji, and L. Vanfretti, "Over-current relay model implementation for real time simulation & Hardware-in-the-Loop (HIL) validation," IECON Proc. (Industrial Electron. Conf., pp. 4789–4796, 2012, doi:10.1109/IECON.2012.6389585.

[43] D. Bian, M. Kuzlu, M. Pipattanasomporn, S. Rahman, and Y. Wu, "Real-time co-simulation platform using OPAL-RT and OPNET for analyzing smart grid performance," IEEE Power Energy Soc. Gen. Meet., vol. 2015-Septe, pp. 1–5, 2015, doi:10.1109/PESGM.2015.7286238.

APPENDIX-A

The detailed parameters of VSC-HVDC test system shown in Figure 7.15, is given below in Table A1. In Table A2, the DC transmission line parameters are specified. The DC transmission line has two nominal π sections, with each section representing a line of length 75 km.

Table A1. Bipolar LCC-HVDC system parameter

Parameters	Rectifier station	Inverter station
Type of system	VSC-HVDC	VSC-HVDC
Power rating	200 MW	200 MW
DC voltage	$\pm$ 100 kV	$\pm$ 100 kV
Current rating	2 kA	2 kA
AC grid	2000 MVA, 230 kV, 50 Hz	2000 MVA, 230 kV, 60 Hz
Transformer ratings	230:100 kV, 200 MVA, 0.15 p.u.	230:100 kV, 200 MVA, 0.15 p.u.
Phase reactor	0.15 p.u.	0.15 p.u.
DC link capacitor	70×10^{-6} F	70×10^{-6} F
Filter 1	40 Mvar, 100 kV, 50 Hz	40 Mvar, 100 kV, 60 Hz
Smoothing reactor	8 mH	8 mH

Table A2. Parameters of nominal-π model of transmission line.

Parameter	Rating
Resistance r Ω/km	1.39×10^{-2}
Inductance l H/km	1.59×10^{-4}
Capacitance c F/km	2.31×10^{-7}
Length of line	75 km
Number of π-sections	2

8

Smart Grid: A Cyber–Physical Infrastructure and Security

Manoj Tripathy, M. Suresh

Electrical Engineering Department, Indian Institute of Technology Roorkee, India
Email: manoj.tripathy@ee.iitr.ac.in, mogilicharla_s@ee.iitr.ac.in

Abstract

For any nation, the energy sector is a vital critical infrastructure, which plays a significant role in its development and economy. A high-quality, reliable, and resilient electricity supply is a necessity in the modern world. To fulfil the requirement, the concept of the smart grid is introduced which is viable due to overwhelming advancement in power electronics converters, smart sensors/meters, and information and communication technology (ICT). This chapter discusses the principle of smart grid-cyber–physical system operation. It deals with the cybersecurity of smart grid infrastructure issues, selection of communication technologies in smart grid, cybersecurity standards, typical type of cyberattack and threats against smart grid, and impact of a cyberattack on smart grid. It explains how a cyberattack handling approach is different from information technology (IT) to operational technology (OT).

Keywords: Cyberattacks, information security, smart grid, information and communication technology

8.1 Introduction

A typical traditional radial electrical system involves bulk centralized generation that has heavy reliance on conventional sources like fossil fuels etc., with limited automation and flexibility in control. This centrally generated power is delivered to the loads via transmission and distribution systems. By the end of the 20th century, the power network had expanded due to high demand for electric power and it became an integrated and complex power grid. The power grid started facing difficulty not only in controlling the long high-voltage transmission lines but also to fulfil the demand for electric supply. Due to the continuous depletion of fossil fuel reserves, environmental pollution, and global warming trends, alternative low-carbon strategies for electricity generation to meet the increasing demand were required. This resulted in integrating renewable energy resources like solar, wind, etc., into the grid. However, due to the intermittent and unpredictable nature of renewable sources, energy fluctuations and uncertainties are introduced into the grid. To handle this situation, the need for an energy storage system like a battery came evident. In recent years, these technologies are heavily being deployed in the distribution grid and thus changing the consumer to act as a prosumer, where customers not only consume power but also supply power back to the grid. To encourage the participation of prosumers, real-time energy marketing has been put in place where the price of electricity is dynamic in nature based on the peak demand. This interaction, management, and flexibility in the operation of the sources requires knowledge of the status of the various sources in relation to each other. This necessity pushed the energy sector into the extensive use of information and communication technology (ICT). The combination of above-mentioned technologies introduced a new concept of the smart grid in which there will be both power flow as well as information flow that is not the case of the traditional grid. Millions of power devices are expected to be completely equipped with two-way, high-speed communication in order to create an interactive, dynamic infrastructure that can handle energy more effectively. The smart grid also includes advanced metering infrastructure (AMI), distribution management system (DMS), microgrid energy management system (MEMS), distributed energy resource management system (DERMS), demand response management system (DRMS), phasor measurement units (PMUs), real-time energy bidding market management system, IEDs, smart sensors, communication networks, and various communication protocols etc., as a part of its supervisory control and data acquisition (SCADA) system. A smart grid, then, can be thought of as a cyber–physical system (CPS) since it consists of a collection of networked agents that interact with the physical world, such as actuators, sensors, control processors, and communication devices. Although this kind of CPS architecture offers visibility and flexibility to the entire operation and control of the smart grid, at the same time this makes it vulnerable to cyber threats. Cyber threats are being referred to as potential sources of asymmetrical warfare in the digital era because of their ability to inflict significant harm with relatively little financial outlay. It is difficult to determine the origin and purpose of such attacks because of the sophistication of the present era, which can hide the real actors behind them. In today's cyber world, terrorists and criminals are targeting critical assets via cyberspace. If those devices are tampered with, it could have a significant impact on the grid. Once the attacker gets access, he/she can modify, delay, or even delete the data, and can issue false control commands etc., that will have serious consequences at the physical level.

Historically, most of the Industrial Control Systems (ICSs) like SCADA are air-gapped, i.e., not connected to the Internet. The operators never thought of issues related to cybersecurity, and their initial intent was simply to have communicable, automated and controllable SCADA systems. In 2010, the Stuxnet worm (the first digital weapon) became the first cybersecurity attack targeting ICSs. Through the use of a USB stick, the malicious programme infiltrated an air-gapped ICS SCADA and surreptitiously accessed Windows computers with the aim of causing physical harm. The infection caused centrifugal pumps to run at dangerously high rates and sustain physical damage by infecting selected programmable logic controllers. To prevent the operator from knowing about the higher speeds (malfunctioning), the attacker recorded sensor data for about 21 s when the system was healthy and when the attack occurred, replayed the same set of data in a loop so that the operator saw normal healthy data (replay attack) [1]. This clearly indicated that keeping ICSs isolated cannot prevent cyberattacks from happening.

Another notable attack on Ukraine's power grid in 2015 was a first-of-its-kind attack that set an ominous precedent for the security of power grids everywhere. Here, the SCADA was connected with IT through a firewall separation. Still the attacker entered the substation SCADA system and tripped about 30 circuit breakers. This was a coordinated cyberattack on three Ukrainian distribution companies, disconnected seven 110 kV and two 335 kV substations which showed the possible scope of a cyberattack by turning off electricity to over 200,000 users. Technically speaking, this attack used firmware modification, spear phishing, and remote-access vulnerabilities to render physical systems unusable and result in significant power outages [2].

In 2016, again another cyberattack was reported on the Ukrainian power grid (Ukraine 2016). In the US, a Utah-based renewable energy provider (sPower) was hit with a cyberattack and lost communication to its power generation installations on March 2019. They lost visibility to their site for two hours and the root cause was the unpatched firewall. A laboratory-level cyberattack demonstrated by the researchers of Idaho National Laboratory (called Arora's attack) exploded the generator. They initiated a test cyberattack that led to the self-destruction of a generator by breaking into its controls and quickly opening and closing its circuits at a rate that was out of sync with the current grid. The generator burst into flames in three minutes [3]. In 2015, 800,000 inverters with a capacity of 154 MW were remotely updated by Enphase Energy on the Hawaiian Islands. This is a breakthrough as it reduces cost by eliminating the need of visiting the site physically and upgrading the firmware. This has raised concerns about cybersecurity. Anyone with access to the control network of a corporation that could remotely alter the settings of hundreds of megawatts of electrical equipment would also be able to make harmful changes to those units [4]. This shows the vulnerability of smart grid which is an ICS and the impacts of a successful attack. If we look at some of the statistics of cyberattacks reported in the critical infrastructure sector-wise, the energy sector is reporting the highest number of incidents than any other amounting to 32%. A survey in 2019 by SysAdmin, Audit, Network and Security (SANS) on the state of OT/ICS cybersecurity survey, revealed that the frequency of cyber incidents reported has increased significantly from 2017 to 2019 [5].

Although people have explored and proposed many technologies for securing information technology (IT) systems, the applicability of the same techniques in the ICS/operational technology (OT) environment like smart grid is challenging. There are some significant differences in the cybersecurity aspects of ICS than that of IT. The security objectives in IT are confidentiality, integrity, and availability (CIA triad) in which confidentiality is given the highest priority and the availability is the least. When it comes to the ICS another extra objective that comes into the picture is safety, given the highest priority and then availability, and last comes integrity and confidentiality. IT security is data-centric whereas ICS is process-centric. In the ICS environment, the ICT hardware and software is meant to detect, monitor, and control the physical devices/processes and events and mostly they might be embedded controllers with limited resources (memory and computational power). These constraints make it difficult to implement cryptography techniques like encryption, hashing, and digital signatures that are available for IT. Moreover, the introduction of these techniques will introduce delay which is not acceptable in ICS due to its time-critical nature (time for sensing, process control decision, and actuation). The diverse nature of

ICS, like different applications require different standards, communication technologies, protocols, time constraints, and criticality levels etc., poses a challenge as a single solution may not be viable for all types of ICS environments. The updating and patching process is not frequent enough as that of IT due to the long lifetime of ICS devices. Also, there are still many legacy systems being used whose vulnerabilities can be easily explored by the attackers and might be publicly available. Also, due to the Always-ON (availability) requirement of the ICS, updating and patching cannot be done simply like IT since it may require a reboot. The impact of a security breach in IT could be information loss or delay in business process. But when it comes to OT which is a cyber–physical system, the impact could be physically catastrophic like physical damage to equipment, human safety, and environmental impact. Another challenge is how to validate and differentiate the dynamic nature of the physical process to be either normal or abnormal state. This is because an abnormal state maybe a result of an actual fault in the system or it could be due to a malicious attack. Further, when we specifically consider the power grid – which is a large cyber–physical system that is geographically dispersed – it is not possible to effectively provide physical security measures to all components. The requirement of being in synchronism at all levels of the grid is another critical factor and involves fine control and balancing [6, 7].

Cybersecurity is one of the most burning problems that industries are facing today. With the technology advances, more things are connected to the Internet and exposed to an attacker. This may involve the critical infrastructure and increase the network or system's attack surface and vulnerabilities. Cyberattack incidents are rising exponentially every year and need attention to develop a defending system against cyberattack.

This chapter briefs about CPS architecture of smart grid, cybersecurity issues, and intrusion detections.

8.2 Background

Smart electric grid is a heterogeneous system that is integrated with smart sensors for measurements at high rates and resolution, distributed renewable energy sources (RESs), loads, communication network, and energy management system for control and protection operation. Interoperability is a key feature of a smart grid which provides capability to communicate, execute programs, or transfer data securely, and effectively between/among two or more systems or devices. To develop interoperability features standards are required and standards development organizations (SDOs) such as National Institute of Standards and Technology (NIST), American National Standards Institute (ANSI), International Electrotechnical Commission (IEC), Institute of Electrical and Electronics Engineers (IEEE), International Organization for Standardization (ISO), International Telecommunication Union (ITU) are working on it [8]. The standards used for smart grid may be classified into five categories as follows:

i. Interconnection of distributed energy resources (DERs) – IEEE 1547, IEC 61850-7-420, IEC-61400-25
ii. Wide area situation awareness (WASA) – IEEE C37.118-2005, IEEE C37.118.1, IEEE C37.118.2, IEC-61850-90-5
iii. Substation protection and automation – IEEE 1379, IEC 61850, IEC/IEEE 60255-24, IEEE C37.240-2014
iv. Time synchronization – IRIG-B, IEEE 1588, IEEE C37.238
v. Cybersecurity – IEC 62351, IEEE 1686, NISTIR-7628 smart grid cybersecurity, IEEE-1547.3 Guide for cybersecurity of DERs, IEC-62443 IACS cybersecurity, NERC critical infrastructure protection (CIP) standard

The protection system applied in the smart grid must respond to both utility grid and microgrid faults. Control systems are probably the most important enabling element of a sustainable smart grid. In a smart grid, there are three different layers of control, i.e., primary, secondary, and tertiary control, as shown in Figure 8.1. Each layer has specific characteristics and performs controlling action under its limits. The primary control is also termed as local control, which manages the power of individual DG and collaborates with other DGs using a secondary layer. Just like primary and secondary control, tertiary control manages power flow between the main grid and microgrid. In order to handle multiple control systems, secondary and tertiary layers are communication-interfaced control. The details of each control layer are as follows:

- *Primary layer control* – Microgrid incorporates different types of DGs, with distinct controlling schemes. Considering the inverter type DG, the individual control scheme is composed of voltage, current, and power control loops. The interconnection of loops manages the power contribution of the single system.
- *Secondary layer control* – Parallel connected DGs have dissimilar power ratings; consequently, the voltage and frequency perturbation are different for individual generating units. In order to achieve synchronism and

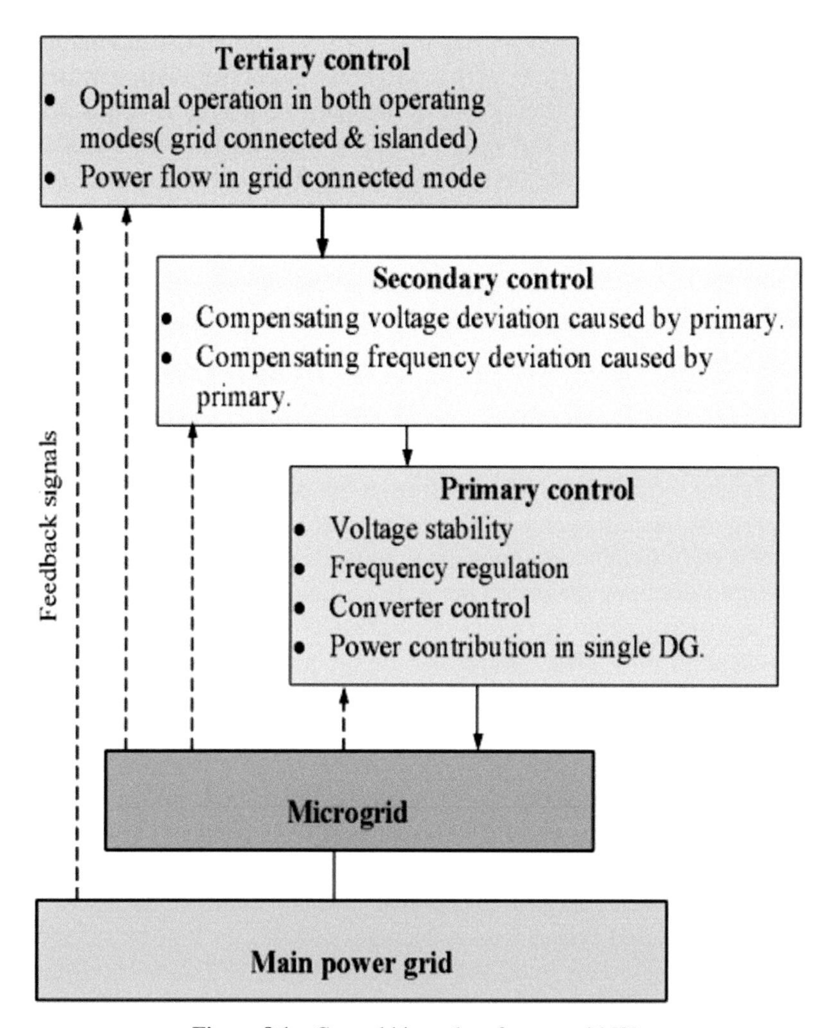

Figure 8.1 Control hierarchy of smart grid [9].

constant voltage at PCC, the secondary control layer has been introduced to the multi-DG microgrid.

- *Tertiary layer control* – It is the topmost layer of the hierarchy, which manages the power flow between multiple grids. It is mainly based on a centralized controller where measured data from different units are acquired, processed in a main controller, and further controlling signals are transmitted to the lower controlling layer using the communication link.

The ultimate goal of a higher control layer is to provide control signals to the primary layer. To enjoy effective automation, control and protection functionalities, the intelligent SCADA is normally deployed in bulk generation plants, transmission and distribution substations for the technical and economic optimization of the grid operation. RESs are integrated with low-voltage distribution networks of a microgrid, the most preferred means of information communication and supervisory control of RESs, storage devices, and loads is a SCADA system. The SCADA systems are used in control centres which use MTUs (master control units) to gather the status data of various circuit breakers, isolators/switches, and IEDs (intelligent electronic devices), and smart meters are connected through RTUs (remote terminal units) via telephone lines, optical fibre cables, or radio channels. At control centres, the information data gathered is displayed on various computer servers for an easy understanding of the operators. I.C. smart grid architectures rely on third-party telecommunication services and heterogeneous systems to fulfil the communication demands of different parties in an inexpensive and efficient manner. Therefore, a smart grid comprises cyber assets such as control, protection, and monitoring devices that are connected via communication link. These points can be the target for an attacker that may be used to cause damage to the physical power system components such as generators, transformers, converters, etc., and hence smart grid is a real cyber–physical system in which ICT needs to be effectively managed and kept secured. In other words, the ICT should be given high priority in the cybersecurity aspects of the smart grid because ICT channels are vulnerable to intrusion via cyberattacks and, thus, jeopardize the operation and control of a smart grid. ICTs are essential for productive and successful communication between the physical and cyber layers, which necessitates the highest security standards. When developing smart grid communication techniques at different levels, such as home area network (HAN), neighbourhood area network (NAN), metropolitan area network (MAN), and wide area network (WAN), two fundamental aspects of communication, namely space and time, which refer to the communication distance and time taken for the transportation of information, should be taken into consideration [10]. The ideal requirements for a smart grid telecommunication system are as follows [11]:

- Be bidirectional
- Be wideband
- Have low latency

- Support massive, real-time data collection from possibly millions of end-points
- Span vast, noncontiguous, and heterogeneous service areas
- Be highly reliable

A telecommunication network is made up of hardware and software that work together to transport data and traffic from origins in service points (SCADA, substations, meters, and so on) to service delivery points – where application servers are located – in a manner that is efficient, scalable, and manageable. Smart grid communication network architecture may be designed at three levels such as core network, access network, and local network. The hub of the network, where all of the components are connected, is known as the core network. Since it is made up of the most crucial network nodes, any failure will have a significant impact on the entire system. It has gigabit transport capacity and is capable of routing high-volume traffic. The wavelength division multiplexing (WDM), synchronous digital hierarchy (SDH), and synchronous optical network (SONET) may be used to interconnect relevant network aggregation points to deal with such a high volume of traffic. An access network serves as a conduit between a local network and a core network. For instance, an access network comprises telecommunication components that link substations, or medium level elements, to each other and to the core network. Additionally, it facilitates the flow of data between central application servers and end devices in this fashion. Since final service locations are dispersed around the region, access networks often incorporate a variety of technologies. In order to incorporate new low voltage (LV) control devices into the smart grid as effectively as possible, local networks are utilized to link substations to field devices like sensors and smart meters.

Another crucial component of a communications network is the planning and construction of the Internet Protocol (IP) network. The scalability and general performance of a network are influenced by IP architecture, which also needs to ensure that network operations are controlled. IP address assignment is a crucial component of IP network design. If local networks are not considered while planning an IP address, IPv4 private ranges are typically sufficient to accommodate smart grid deployments. Although there is a chance that IPv6 may be deployed in smart grid networks, its widespread adoption does not mean that it will be a common option if the addressing space needs to be expanded [11]. Network security is a component of network design decisions and needs to be considered in a comprehensive manner to account for any issue that may impact the privacy, confidentiality, integrity, or access of any information related to the smart grid.

An open system is a collection of protocols that permits communication between any two systems without requiring modifications to the underlying hardware and software's logic. The open systems interconnection (OSI) model, as shown in Figure 8.2, is a layered framework to design a network system that establishes communication between different types of communicating devices [12]. First, second, and third layers i.e., physical, data link, and network layers are the communication network support layers which focus on the physical connection, physical addressing, and transport timing. These layers deal with physical aspects of moving data from one device to another device. And fifth, sixth, and seventh layers i.e., session, presentation, and application layers allow interoperability among unrelated software systems. At each layer, a header or trailer bit can be added to the data unit. Table 8.1 demonstrates generalized security services according to the OSI model.

Figure 8.3 illustrates the list of attacks that may occur at different OSI reference model; in other words, it may be said that the cyberattack mapped the OSI layer for the ICS and IT systems respectively.

In a smart grid, different types of communication protocols are utilized and each communication system has different requirements like latency,

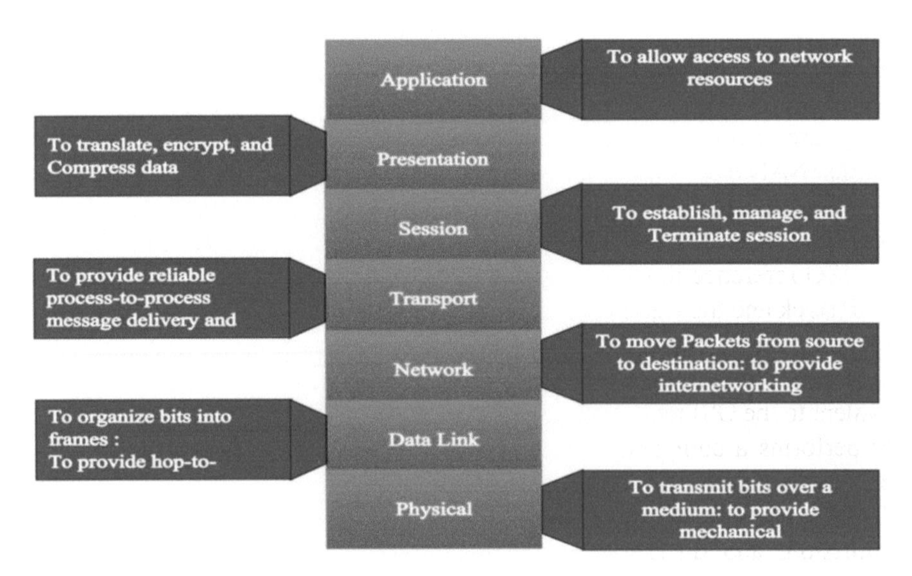

Figure 8.2 Seven layers of OSI model and its function.

Table 8.1 Security services according to OSI model.

Layer number	Layer	Service
7	Application	Authentication Access control Data integrity Data confidentiality Nonrepudiation
6	Presentation	Data confidentiality
5	Session	-
4	Transport	Authentication Access control Data integrity Data confidentiality
3	Network	Authentication Access control Data integrity Data confidentiality
2	Data link	Authentication Access control Data integrity Data confidentiality
1	Physical	Data confidentiality

security, and packet size. Generally, communication protocols like IEEE 802 series, Modbus, DNP3.0, ASCII, IEEE C37.118, IEC 60870, IEC 61850, ANCICI12.18, C12.19, C12.22, IEC62056, OSGP etc., are used in smart grid communication network. The IEEE802 series was created to facilitate the use of home area networks (HANs) in customer premises, neighbourhood area networks (NANs) in distribution networks, and local area networks (LANs) in SCADA. Ethernet, wireless LANs, Bluetooth, Zigbee, and other technologies are covered by the IEEE 802 series, which is based on the lowest two layers of the ISO reference model. Figure 8.4 illustrates the mapping between the TCP/IP stack and the OSI model stack. Typically, application-layer protocols that operate on a TCP/IP network are used by smart grid devices to interact [13]. As illustrated in Figure 8.4, the TCP/IP stack's network interface layer is equivalent to the OSI model's physical and data link layers, while the Internet layer performs a comparable function to that of the OSI model's network layer. The transport layer, which is shared by both models, comes next.

The functionality of the power grid is broadly divided into generation, transmission, and distribution and each functional division has a dedicated system for controlling a specific device/machine by using a specific telecommunication system. As given in Table 8.2, Govindarasu et al. reported

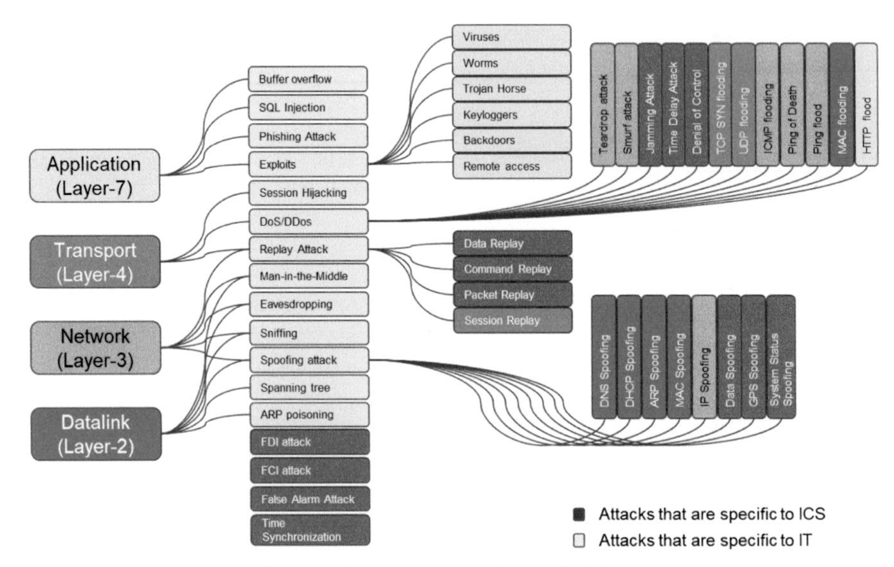

Figure 8.3 Attack mapping to OSI layer.

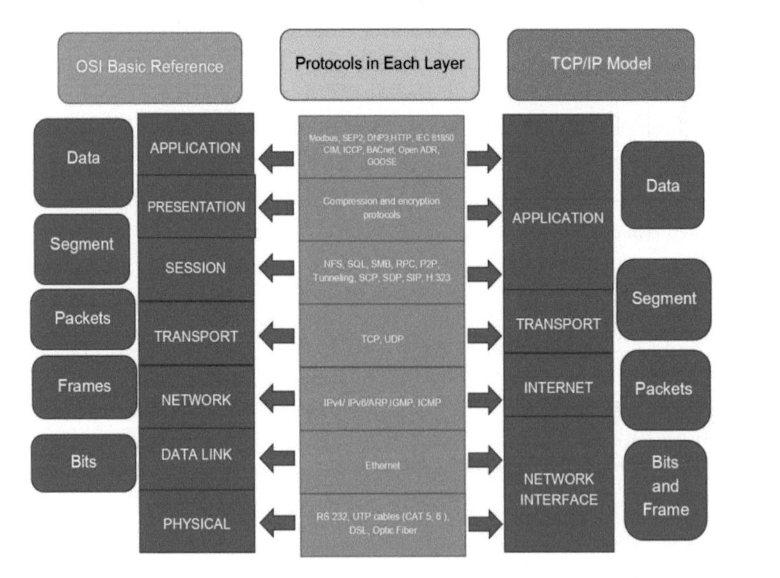

Figure 8.4 Logical mapping between OSI basic reference model and TCP/IP stack [13].

the taxonomy of control loops of a power grid [14]. From Table 8.2, it is clear that all three sections i.e., generation, transmission, and distribution are cyber vulnerable.

Table 8.2 Smart grid – cyber–physical systems control taxonomy [14].

Domain	Control	Control attributes						
Power grid		1 Physical parameter	2 Measurements and inputs	Communication messages 34		5 Computation	6 Machine/device	7 Control action
				Data acquisition	Messages control			
Generation	*Automatic voltage regulator*	Terminal voltage	Measured and reference terminal voltage	Local measurement from terminal	Local message to excitor control	Calculation of excitation current	Generators	Increase/decrease excitor current
	Governor control	Rotor speed	Measured and reference rotor speed	Local measurement from rotor speed sensor	Local message to prime mover controller	Valve position	Prime mover	Open/close valve
	Automatic generation control	Frequency	Frequency and tie line power measurement	Wide area communication (IEC 61850)	Point-to-point communication (DNP 3.0)	Area control error (ACE) calculation	Generators	Raise/lower generation
	Security constrained economic dispatch	Power generation	Demand, network topology and line limits	Wide area communication (IEC 61850)	Point-to-point communication (DNP 3.0)	Generation set points	Generators	Generation re-dispatch
Transmission	*State estimation*	Power generation and network topology	Voltage and power, VAR or current flow	Wide area communication (IEC 61850)	Point-to-point to switchyards and generating stations	System voltage and phase angle calculation	Generators and switching devices	Generation re- dispatch and open/close breakers
	VAR compensation	Voltage	Reference voltage, measured voltage and VAR device parameters	Local measurement	Local message to FACTS device	Reactive power level calculation	FACTS	Absorb/supply reactive power

Table 8.2 Smart grid – cyber–physical systems control taxonomy [14].

Domain	Control	Control attributes						
	HVDC transmission control	DC voltage and current	Reference voltage and measured voltage	Local measurement of voltage	Local message to converters	Firing angle	Power electronic converters	Increase/decrease firing angle
Distribution	***Demand side management***	Load scheduling	Demand, conventional and alternate resources availability	Power demand request	Allotted schedule to factories and homes	Load schedule computation	Loads	Turn on/off load
	Load shedding	Load connected to system	Generation limit, system frequency and current generation	Local frequency measurement and generation level from control centre	Trip message to relays on distribution feeder	Load amount and location	Distribution feeder	Open feeder breaker
	Advanced metering infrastructure	Consumer load	MDMS/headend instructions	NA	Disable/load shed	Meter function	Consumer meter	Disable meter/shed load

8.3 Cybersecurity Issues in Smart Grid

A typical architectural model of a smart grid network (SGN) proposed by the National Institute of Standards and Technology (NIST) is shown in Figure 8.5 [15].

This conceptual model will give a brief overview of the actors in a smart grid and the interactions among them in terms of power flow as well as the information flow for a successful operation of a smart grid. This has majorly seven domains namely generation, transmission, distribution, operations, energy market, customer and service provider that are interacting with each other. Typically, the power flow will be from generation to transmission to distribution to customer and thus mostly among four domains. At the same time if we observe the information flow, it is among all the seven domains and thus we can understand that the smart grid operation is heavily dependent on the use of ICT. This involves the use of network components, TCP/IP/Ethernet-based protocols, personal computers, Microsoft Windows,

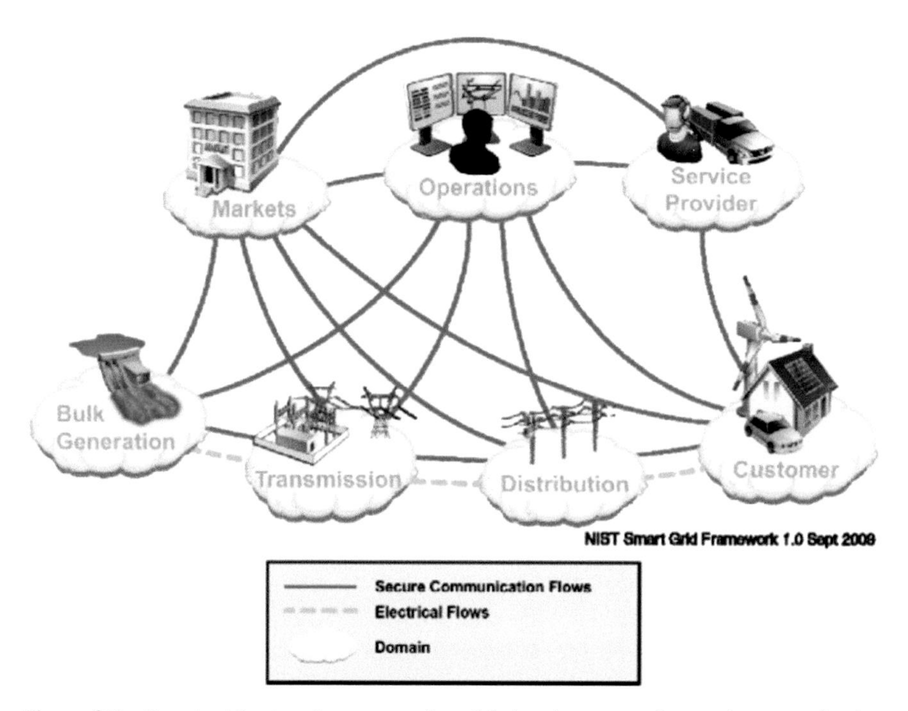

Figure 8.5 Smart grid network conceptual model showing power flow and communication interactions [15].

communication networks like HAN, field area network (FAN), WAN, etc., thus expanding the attack surface and increasing threat of cyberattack. Depending on the actors and the interactions happening between them, the attacker can target multiple objectives to disrupt the operation of the grid. In a smart grid, attackers may target sensor devices, IEDs/relays, actuators, computing components, communications and feedback systems/controllers. Few possible attack scenarios in the above smart grid domains are briefed as follows [16–17].

Automatic generation and control (AGC): The AGC will function to adjust the output power of site/plant generators to ensure the power balance, grid frequency within acceptable limits and the power exchange in an interconnected grid as per the scheduled values. This will require various measurements like tie-line power, frequency etc., as input information and will adjust the governor output as per the control logic. This usually runs and issues control signals in a time scale of seconds, and thus it may not be possible to employ data validation techniques in a legacy control system. An attacker can take the advantage of this to manipulate the measurement inputs to the control loop of AGC without being detected. For example, if the attacker could gain access to the $P_{tie-line}$ and frequency measurements, he/she can introduce false measurement data to fool the AGC to believe that there is a surplus/deficit in the generation which may not be true. Then AGC will send the commands to ramp down/ramp up the generation in which either of the cases leads to a power imbalance and frequency deviations. If the frequency deviation exceeds the pre-set limits, then the protection relay system will act and trip the entire plant generation which will worsen the scenario. An attack on AGC would have a direct impact on the system stability, frequency and thus the motive of the attacker is to cause system wide stability issues in the grid.

State estimation: It is a steady-state process that evaluates the system variable such as bus voltage and its phase angle. This process requires various measurements data like power injections, voltages, line flows, and status of circuit breakers from different locations of the grid. Based on this state estimation result, the control centre will take the decisions like load shedding, ramping of generation, etc. Usually, these measurements are communicated to the control centre via WAN, thus there is a wide attack surface available for the attacker to enter. An attacker can target on measurements and introduce false data injection attacks to mislead the state estimation process. Further, he/she can also modify the status information of the circuit breakers and thus could fake a line outage. This happens at the transmission level.

Advanced metering infrastructure: When it comes to the distribution system, advanced metering infrastructure like smart meters will play an important role in the smart grid. These smart meters are deployed at the consumer's location to provide more visibility to the level of consumer load consumption. This can also be used for load shedding/disconnecting consumer appliances during peak load times. Further, it can be used for smart billing systems, prepaid energy billing schemes, etc. These multiple interactions can provide opportunity to the attacker with a wide variety of attack vectors. For example, he/she can manipulate the meter reading to affect the billing operation, issue connect/disconnect control signals to affect the consumer load, etc.

Energy market: This domain deals with the real-time pricing of the electricity as per the demand by exchanging information of real-time pricing signals with the consumers/prosumers to participate in the real-time market/energy bidding. The motive of the attacker is to create economic impact on the system. For example, a denial-of-service (DoS) attack on the pricing information jams the signals to a group of selected users. As they are unaware of the change in the price signal, they might be taking their consumption decisions based on the old price signals. Moreover, there is a possibility to introduce false pricing information into the grid that will misguide the consumer's decision on their consumption.

Therefore, the main concerns of the smart grid's cybersecurity are information protection, infrastructure protection, and application security. Information protection includes confidentiality, integrity, availability, and authentication of data. Infrastructure protection focuses on security of communication infrastructures like routers, DNS servers, links, and Internet Protocols. Whereas application security ensures the safety of generation control applications, transmission control applications, and distribution control applications.

8.4 Intrusion Detections

Cybersecurity of smart grid is critical as many deliberated cyberattacks and random failures may occur. In order to detect, prevent, and mitigate such possible attacks intrusion detection systems (IDSs) are developed as the main tool. An IDS is an autonomous process that finds events of violation of security policies or standard security practices in communication networks. IDS is one of the most common tools for a security infrastructure which is located inside the network to monitor all internal traffic. The performance of

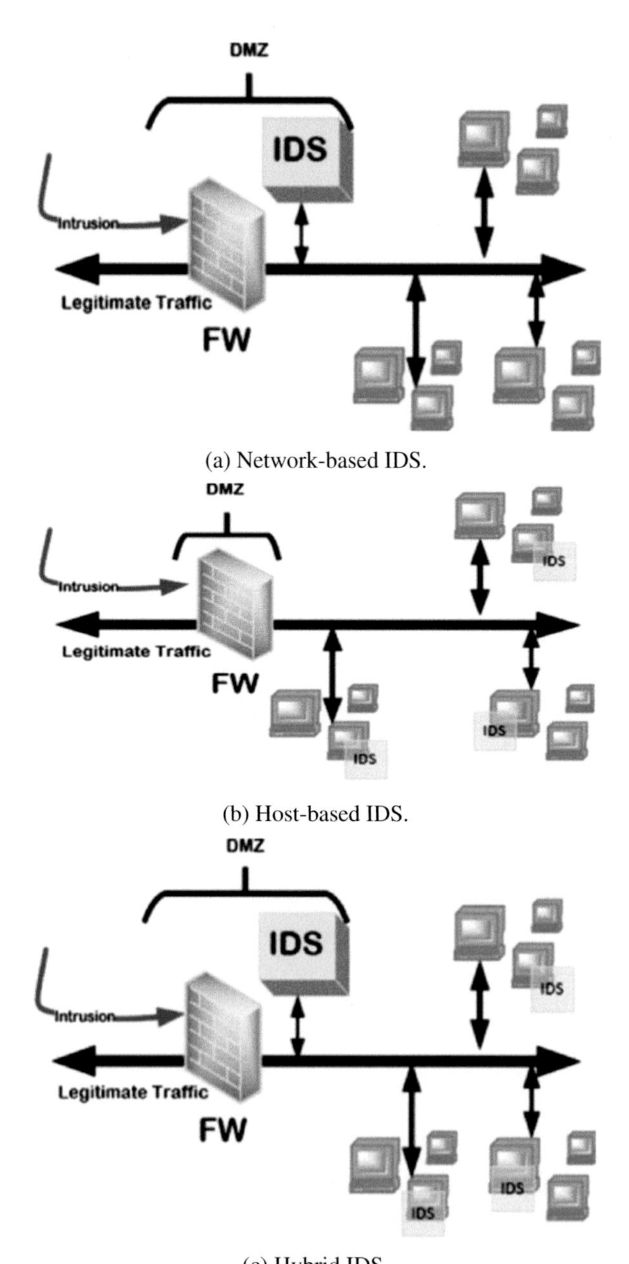

(a) Network-based IDS.

(b) Host-based IDS.

(c) Hybrid IDS.

Figure 8.6 Typical example of (a) network-based IDS, (b) host-based IDS, (c) hybrid IDS [18]

an IDS is judged based on the detection rate (DR), the lowest false alarm rate, and the ability to catch unknown attacks. IDS may be classified into three classes: network-based, host-based, and hybrid-based IDS as shown in Figure 8.6 [18]. In network-based IDS, an IDS module is placed inside the network that can check the whole network's malicious activities by inspecting all packets moving across the network. The network-based intrusion detection system (IDS) relies on known assaults and is maintained by frequent software upgrades and patching. It also dynamically monitors network traffic to identify and stop unknown and suspicious attacks. As seen in Figure 8.6 (b), host-based IDSs maintain an IDS module on each client connected to the network, monitoring all incoming and outgoing traffic for that client. Host-based IDS is used to monitor the activities, log events of a host system along with its network traffic, and check if a system has been attacked or not. It is further divided into four subsystems: kernel-based intrusion detection, connection analysis, log file analysis, and files monitoring system. A file monitoring system routinely checks and compares files that hold data on a machine's size, modifying history, and user files with previously collected files. These edit modifications can be identified in the event that an attacker gains access to the host's computer and attempts to edit files. The logged file is examined in log file analysis to see if unusual data and activities are recorded or not. The connection analysis detects TCP/IP packets on the network, considers the erroneous sequence of TCP packets arriving at the host computer, and looks for evidence of intrusion activities. An additional element to improve kernel security is kernel-based host intrusion detection system (IDS), which allows the kernel to identify and stop hostile activity and unusual behaviour [19]. But in case of hybrid IDS as shown in Figure 8.6 (c), IDS modules are in the network as well as at each client to monitor both specific client and network activities at the same time. IDS can be classified into three types based on the detection method: misuse/signature, anomaly, and stateful protocol/specification-based IDS. An intrusion detection system (IDS) that is based on misuse/signatures may identify any type of attack by comparing its attributes to previously stored signatures or attack patterns. While known attacks can be detected with this kind of IDS, it is harder to identify unknown or novel assaults. Anomaly/behaviour-based IDS utilizes the knowledge of data base that contain profiles, patterns of user/data. It takes input from the audit records generated from the operating system. It may consider the static information like structural and syntax features of the application/programmes or it may take dynamic information like run behaviour and dynamic analysis of structural and syntax features or it may combine both static and dynamic information to create behaviour/patterns. A manually defined set of rules and constraints is used by a specification-based intrusion detection system (IDS) to express regular operations; any divergence from these parameters during execution is deemed malicious [20]. Table 8.3 represents a comparative study of IDS types based on the methodology.

In cyber–physical system attack, a tree is used to analyse potential threats as shown in Figure 8.7 [21] and the concept of an attack tree is applicable to cybersecurity of smart grid too. The tree branches include the following types of attacks such as sensor devices attacks, attacks on actuators, attacks on computing components, attacks on communication, and attacks on feedback. In smart grid, cyberattack mainly targets on availability of services, integrity, and confidentiality. Denial-of-service (DoS) attacks, attempt to delay, blocking or corrupting the communication channel in the smart grid are typical examples of cyberattack on availability of services. DoS may be done at different OSI layers/forms such as at physical layer by channel jamming, DoS attack can be done. At MAC layer deliberately modifying the MAC parameters (backoff parameters), a DoS attack and spoofing attack are possible. Due to the possibility of distributed traffic flooding and worm propagation attacks on the Internet, the network and transport layers are susceptible to denial of service (DoS) attacks. The integrity attack is done by deliberately and illegally modifying or disrupting data exchange between two devices of the smart grid. In case of confidentiality, aim is to acquire unauthorized information from network resources in the smart grid.

The cyberattacks immediately affect the availability of communication systems, control systems, and protection systems which are vital security threats in smart grid. Cyberattack mitigation mechanisms can be deployed

Table 8.3 Comparison of IDS types based on the methodology [18].

Types of IDS	**Misuse/signature based**	**Anomaly/behaviour based**	**Specification based**
IDS methods	Identify known attack patterns	Identify unusual activity patterns	Identify violation of predefined rules
Detection rate	High	Low	High
False alarm rate	Low	High	Low
Unknown attack detection	Incapable	Capable	Incapable
Drawback	Updating signatures is burdensome	Computing any machine learning is heavy	Relying on expert knowledge during defining rules is undesirable

Figure 8.7 A tree diagram of attacks and threats on cyber–physical systems [21].

in network according to the concerns of the smart grid's cybersecurity such as information security, infrastructure security, and application security. For information security, digital signature, message authentication code, and public key infrastructure may be utilized. Firewalls, intrusion detection, secure protocols, authentication protocols, attack attribution, and secure servers etc., are used for infrastructure security. And for application security, techniques like risk modelling and mitigation are applied, along with attack-resistant control algorithms and model-based algorithms (such as anomaly detection, intrusion tolerance, and bad data removal).

8.5 Conclusion

The security element of the cyber–physical system perspective of smart grids is a relatively new field of study that is gaining rapid interest from the government, business community, and academic community. The concept of the cyber–physical system of the smart grid, security concerns, the communication architecture, security requirements, security vulnerabilities, intrusion detection, and strategies for preventing and defending against cyberattacks were all covered in this chapter. Additionally, it is summed up that in order to ensure effective and safe information transport in the smart grid, standard and secure network protocols must be used. In this chapter, it is clearly explained how the cyberattack handling approach is different from information technology (IT) to operational technology (OT) system. Furthermore, it is concluded that both information security and system theory–based security are essential to securing cyber–physical smart grid system. It will give new sight to electrical engineers to see the electric grid and help them to make secure the electric grid.

References

[1] A. S. Musleh, G. Chen and Z. Y. Dong, "A Survey on the Detection Algorithms for False Data Injection Attacks in Smart Grids," IEEE Transactions on Smart Grid, vol. 11, no. 3, pp. 2218-2234, May 2020.

[2] V. L. Do, L. Fillatre, I. Nikiforov and P. Willett, "Security of SCADA systems against cyber–physical attacks," IEEE Aerospace and Electronic Systems Magazine, vol. 32, no. 5, pp. 28-45, May 2017.

[3] Z. Li, M. Shahidehpour and F. Aminifar, "Cybersecurity in Distributed Power Systems," IEEE Proceedings, vol. 105, no. 7, pp. 1367-1388, July 2017.

[4] Lai, Christine & Jacobs, et al, "Cyber Security Primer for DER Vendors, Aggregators, and Grid Operators," Tech. Rep., Sandia National Laboratories, Albuquerque, USA, Dec. 2017. [online]. Available: https://doi.org/10.2172/1761987.

[5] Barbara Filkins and Doug Wylie, "SANS 2019 State of OT/ICS Cybersecurity Survey," Tech. Report by SANS June 2019.

[6] Quick Start Guide: An Overview of ISA/IEC 62443 Standards Security of Industrial Automation and Control Systems.

[7] NIST SP 800-82 revision 2, Guide to Industrial Control Systems (ICS) security.

[8] T. Sato et. al. "Smart Grid Standards Specifications, Requirements, And Technologies" John Wiley & Sons Singapore Pte. Ltd., 2015.

[9] A. Bidram, A. Davoudi, "Hierarchical Structure of Microgrids Control System", IEEE Transactions on Smart Grid, vol. 3, no. 4, pp. 1963-1976, May, 2012.

[10] X. Yu, Y. Xue "Smart Grids: A Cyber–Physical Systems Perspective", Proceedings of the IEEE, vol. 104, no.5, pp. 1058-1070, May 2016. DOI: 10.1109/JPROC.2015.2503119

[11] A. Sendin, et.al. "Telecommunication Networks for the Smart Grid", Artech House, 2016. ISBN 13: 978-1-63081-046-7

[12] B. A. Forouzan, "Data Communications and Networking", 4th Edition, Tata McGraw-Hill Publishing Company Ltd., 2006.

[13] A. Sundararajan, A. Chavan, D. Saleem, A. I. Sarwat, "A Survey of Protocol-Level Challenges and Solutions for Distributed Energy Resource Cyber-Physical Security", Energies, vol.5, no. 40, pp.1-18, 2016.

[14] M. Govindarasu, A. Hann, P. Sauer "Cyber-Physical Systems Security for Smart Grid" White Paper, Power Systems Engineering Research Center, Iowa State University, Department of Electrical and Computer Engineering, 3227 Coover Hall Ames, IA 50011, USA, February 2012.

[15] G. Locke, P. D. Gallagher, "NIST Framework and Roadmap for Smart Grid Interoperability Standards, Release 1.0", U.S. Department of Commerce, January 2010.

[16] K. Chatterjee, V. Padmini and S. A. Khaparde, "Review of cyber attacks on power system operations," IEEE Region10 Symposium (TENSYMP), pp. 1-6, Cochin, India, 14-16 July 2017. DOI: 10.1109/TENCON Spring.2017.8070085.

[17] S. Sridhar, A. Hahn, M. Govindarasu "Cyber-Physical System Security for the Electric Power Grid," IEEE Proceedings, vol.100, no.1, pp.210-224, 2012.

[18] K. Kim, M. E. Aminanto, H.C Tanuwidjaja, "Network Intrusion Detection using Deep Learning: A Feature Learning Approach", Springer Singapore, Springer Nature Singapore Pte. Ltd., 2018. https://doi.org/10.1007/978-981-13-1444-5.

[19] K. Letou, D. Devi, Y. J. Singh, "Host-based Intrusion Detection and Prevention System", International Journal of Computer Applications, vol. 69, no.26, pp.27-33, May 2013.

[20] R. Mitchell, I. R. Chen, "Behavior rule specification-based intrusion detection for safety critical medical cyber physical systems," IEEE Trans. Dependable Secure Comput., vol. 12, no. 1, pp. 16–30, Jan 2015.

[21] R. Alguliyev, Y. Imamverdiyev, L. Sukhostat, "Cyber-physical systems and their security issues", Computers in Industry, vol.100, pp. 212–223, 2018.

9

Protection Challenges During Power Swing and Soft Computing Solutions

Jitendra Kumar

NIT Jamshedpur, India
Email: jitendra.ee@nitjsr.ac.in

Abstract

Power swing is the condition where three-phase power varies with advancing and retarding of power angles of the buses in the transmission network. This change in power angle can occur due to a sudden load change, line switching, and/or generator disconnection, etc. Variation in power angle causes additional frequency components to be introduced in the system. Due to this phenomenon, distance relay may malfunction as the power swing is occurring. To prevent such malfunctioning, the distance relay is blocked by the blocking signal generated by a power swing blocker (PSB). During this period, the transmission system remains unprotected, which is undesirable. Under such circumstances, an additional relay can be utilized with suitably designed protection algorithms to protect transmission system. If the line is series compensated, then the protection challenge becomes more complex during the power swing. In this chapter, fault detection and fault direction estimation issues have been discussed. The combination of these two issues makes the fault section identification, which is also deliberated in this chapter.

Keywords: Power swing, fault detection, direction estimation, fault section identification

9.1 Introduction

A power swing occurs in a transmission system due to sudden changes such as a load disconnection, line outage, or fault, etc. Due to this, lower-frequency components are added with fundamental frequency components, and together these can be responsible for the unreliable operation of protection systems. In such cases, relay algorithms may provide incorrect information about the fault or disturbance. During a power swing, a distance relay may maloperate and take an unwanted trip decision as the impedance trajectory comes into the relay zone. To prevent such a situation from arising, power swing blocker (PSB) generates a signal to block the distance relay element. During such operations, the distance relay becomes silent, and the transmission system remains unprotected.

These protection issues can become more complex if the line is series compensated since power angle is also related to series compensation. Therefore, the impact of series compensation on the power swing is also important to analyse and discuss in this chapter. For proper and reliable operation, an additional relay is attached with the distance relay which may take a decision regarding the fault during the power swing.

Relay operates based on the internal algorithms executed according to the data analysis by the soft computing algorithm. Those soft computing algorithms calculate the index/indices using 3-phase voltage and current data measured at relay location and these index/indices are compared with set threshold to get the various information regarding faults. In series compensated line also, voltage can be measured either on bus side (before series compensation device) or online side (after series compensation). Line current can be measured on bus side or line side because it remains same for a line. These voltage and current data is utilized by algorithms of soft computing to analysis of fault and other security purpose also.

9.2 Impact of Power Swing on Conventional Distance Protection Including Series Compensation

Series compensation of a line increases the power transfer capability, stability etc., in a power system. But also, series compensation affects the distance relay operation. Due to sudden changes in the power system, there is a mismatch created between mechanical and electrical power. Due to this mismatch, the load angle varies and may not come into equilibrium resulting in oscillations in the system voltage and current.

For basic analysis, the simple two-source system described in Figure 9.1 is considered. This simple system is well known in the literature [1] and has source impedances Zs1 and Zs2 connected through a transmission line of impedance Z_L. The impedance measured at the relay Z_r is derived in eqn

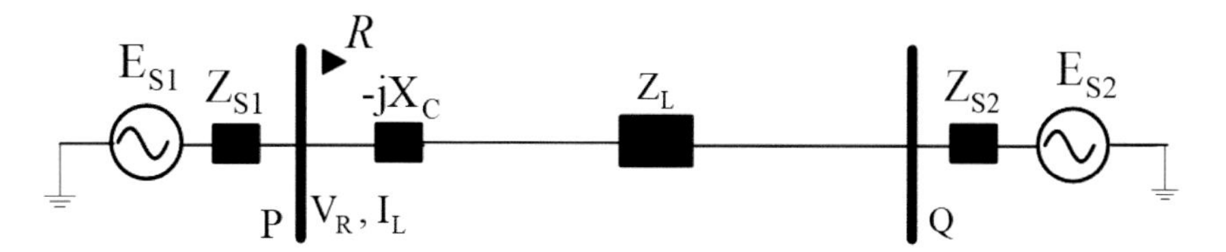

Figure 9.1 Two-source power system.

(9.1). The relay R is positioned at bus-P and used to measure the three-phase voltages and currents. jX_c is the reactance offered by the series compensation device. Here, $Es1$ and $Es2$ are the two voltage sources connected at bus-P and bus-Q, respectively as shown in Figure 9.1. Z_r is the ratio of measured voltage (V_r) and current (I_L) measured at bus-P and can be represented in eqn (9.1).

The current measured at relay location measured can be presented as

$$I_L = \frac{E_{s1} - E_{s2}}{Z_{s1} + Z_L + Z_{s2} - jX_c} = \frac{E_{s1} - E_{s2}}{Z_T}, \tag{9.1}$$

where $Z_T = Z_{s1} + Z_L + Z_{s2} - jX_c$.

If $E_{s1} = |E_{s1}| \angle\delta, \quad E_{s2} = |E_{s2}| \angle 0, \text{ and } h = \left|\frac{E_{s1}}{E_{s2}}\right|$.

Impedance measured at relay location can be derived as

$$Z_r = \frac{V_r}{I_L} = \frac{he^{j\delta}}{he^{j\delta} - 1} \times Z_T - Z_{s1}. \tag{9.2}$$

Here, δ is the power angle.

For $h = 1$, the impedance (Z_r) is presented as follows:

$$Z_r = \left(\frac{Z_T}{2} - Z_{s1}\right) - j\frac{Z_T}{2}\cot\frac{\delta}{2}. \tag{9.3}$$

Eqn (9.3) shows that the impedance measured by the relay is dependent on the power angle (δ) and magnitude ratio (h). Due to variation of δ caused by a sudden change in the system power swing occurs and relay impedance (Z_r) varies. During a power swing, the trajectory of the impedance Z_r (measured at the relay) follows the path when the angle difference of the two source voltages varies and follows a straight line (SO) as shown in Figure 9.2 and can also be analysed in Figure 9.3 for various voltage ratios. In Figure 9.2, “o” point is known as the electrical centre of the swing. This may vary according to the compensation level. The line segments SS1 and SS2 meet at the point S.

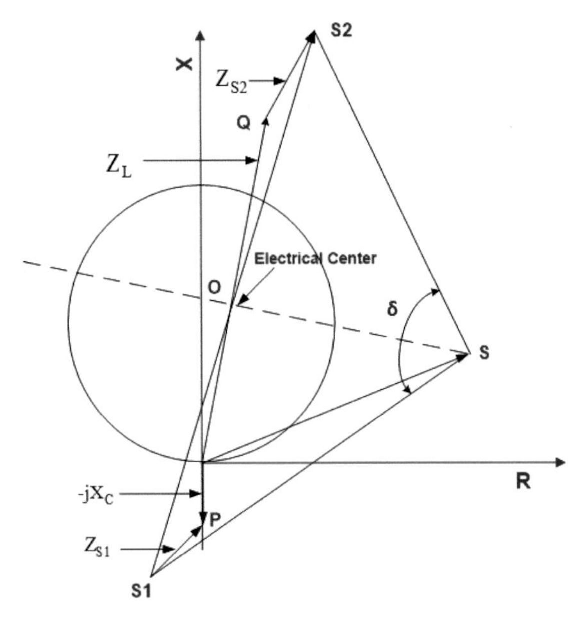

Figure 9.2 Trajectories of impedance of mho distance relay during a power swing.

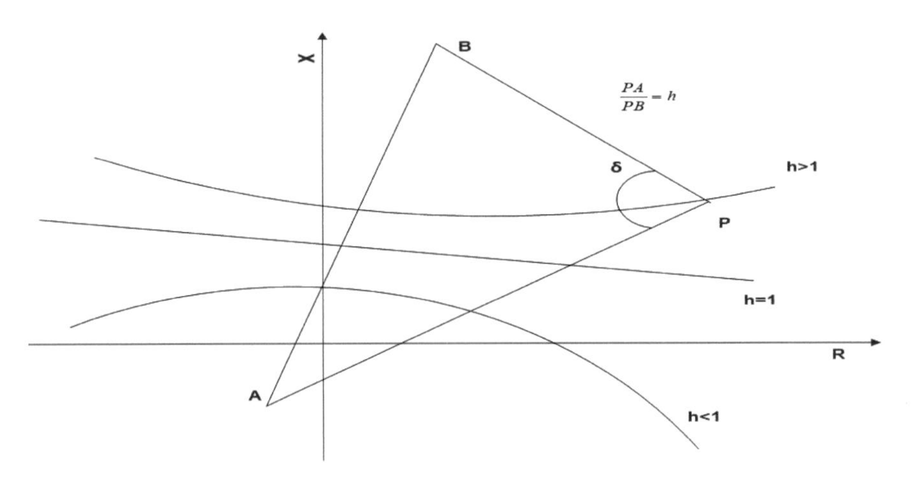

Figure 9.3 Trajectories of impedance at the relay within period of a power swing for various values of *h*.

The angle between the two segments SS1 and SS2 is δ. When the value of this angle δ approaches 180 degrees, the impedance is exactly at the position of the electrical centre. The trajectory path of impedance may differ for h less than, equal to, or greater than one as shown in Figure 9.3. During a power swing, this angle δ moves towards the relay zone. Its movement is slow during the power swing, but its movement is relatively fast during a fault condition.

9.3 Detailed Analysis of Effect of Series Compensation Amid Power Swing

When a series compensated line is employed in the power system, it enhances the power transfer capacity due to reduction of reactance of the transmission line. It also improves the stability of the system. The control action of thyristor-controlled series compensator (TCSC) is also affected by the power swing. However, series compensation is directly related to the power angle and reduces the impact of a power swing. This is analysed further in this section.

The connection associated with initial rotor angle (δ_0) and reactance offered by the series compensation device X_{TCSC} (Xc) can be written as

$$\delta_0 = \sin^{-1}\left(\frac{P_{em}}{E_{s1}E_{s2}}\left(X_L - X_C\right)\right), \tag{9.4}$$

where resistances of the sources and transmission line are considered small and neglected as compared to their reactance X_L. P_{em} is the transfer capacity of the transmission line. Initial rotor angle (δ_0) is dependent on the series compensated reactance.

As a function of rotor angle and X_C, the electrical power output is expressed as

$$P_{em} = \frac{EV}{X_L - X_C}\sin\delta. \tag{9.5}$$

In case of a fixed compensation level of X_C, ΔP_{em} is proportional to $\Delta\delta$. Variation in compensation level of X_C leads to variation in the rotor angle.

If the rotor angle, power output, and compensation level deviations are denoted by $\Delta\delta$, ΔP_e, and ΔX_C, respectively, then eqn (9.5) is expressed as

$$P_{em} + \Delta P_{em} = \frac{EV}{X_L - (X_C + \Delta X_C)}\sin(\delta + \Delta\delta). \tag{9.6}$$

After assuming some trigonometric approximations [26], the eqn (9.4) can be rearranged as

$$\Delta\delta = \frac{1}{P_s}\left[\left(1 - \frac{X_C}{X_L}\right)\Delta P_{em} - \left(\frac{P_{em}}{X_L}\right)\Delta X_C - \left(\frac{1}{X_L}\right)\Delta P_{em}.\Delta X_C\right], \quad (9.7)$$

where synchronizing power $P_s = P_{max}\cos\delta_0$.

In eqn (9.7), it is evident that for the fixed compensation (ΔX_C = 0), ΔP_{em} depends on changes in rotor angle. On the other hand, when variable compensation is used, rotor angle changes are contingent on negative values of ΔX_C. The explanation above suggests that a series adjusted

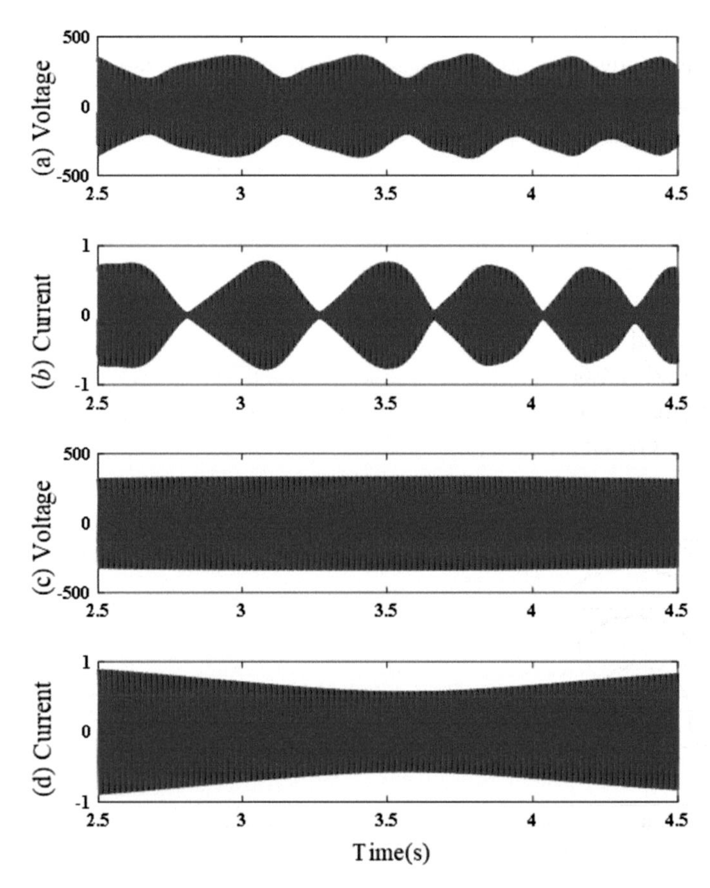

Figure 9.4 Relay signals during power swing in the case of TCSC compensation: (a) voltage with 25% compensation, (b) voltage with 75% compensation, (c) voltage with 25% compensation, (d) current with 75% compensation.

line's increased compensation level reduces rotor oscillation and is also responsible for the power swing's decreased intensity. To support this reasoning, Figure 9.4 displays voltage and current charts for 25% and 75% series compensation in line. Figure 9.4 shows that when 75% compensation is used (Figure 9.4(c,d)), the voltage and current signals exhibit a lower intensity of power swing than when 25% compensation is used (Figure 9.4(a,b)).

9.4 Protection Schemes Against Faults Amid Power Swing

During a power swing, the phasor calculation does not provide the stable magnitude as voltage and current data varies with additional frequency components along with fundamental frequency components. Due to these additional frequency components, the transmission system becomes unstable, and three-phase power also oscillates. High oscillations are responsible for an unstable power swing, and low oscillations lead to a stable power swing.

In the duration of stable power swing, the trajectory of impedance enters in zone-3 and after some time it returns, as shown in Figure 9.5(a). Under such a situation, unintended trips may take place. A PSB element blocks the distance relay during such a period to prevent an unintended trip. But, during this period, the system becomes unprotected from faults. Voltage and current data are procured after simulation of WSCC nine-bus system [2] for a two phase-to-ground (BCG type) fault incepted at 3 s during a power swing and shown in Figure 9.6. Impedance trajectory during fault immediately moves to the impedance characteristics of zone-1 as shown in Figure 9.5(b) if there is

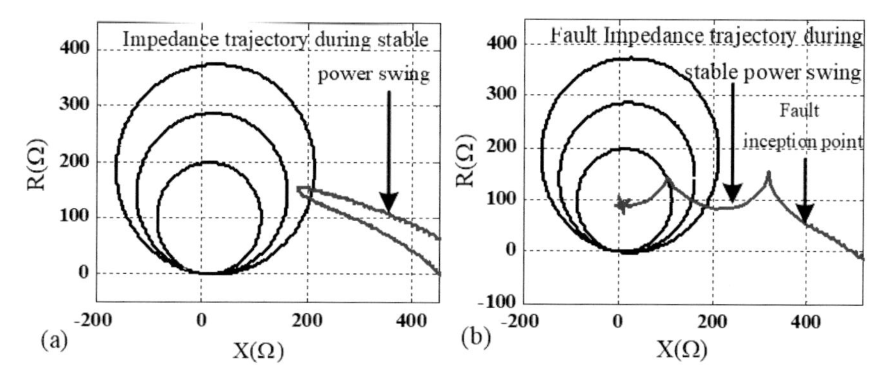

Figure 9.5 Impedance trajectories (a) for stable power swing (b) for bg-type fault during stable power swing.

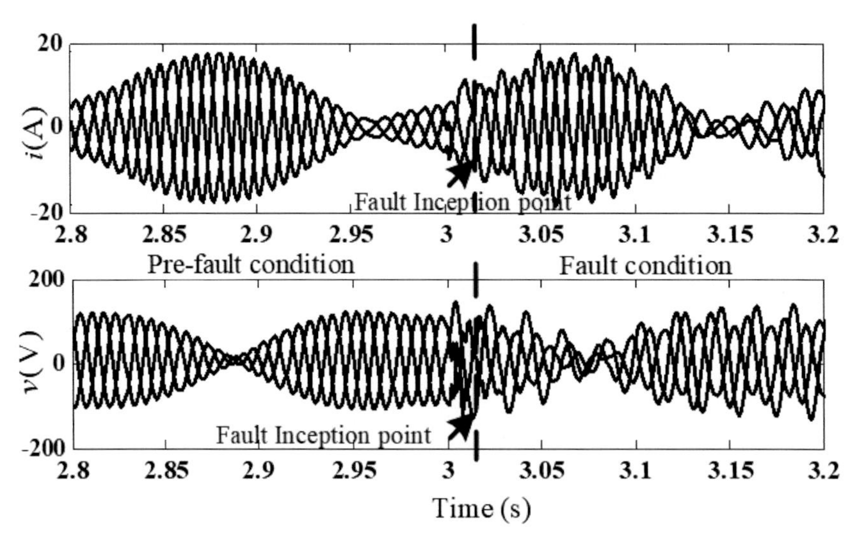

Figure 9.6 3-phase voltages and currents in the power swing duration.

no PSB. Now, there is a requirement of an additional relay which can protect the system during the PSB blocking period.

The concept of an additional relay is implemented when the main relay is blocked for any reason. This additional relay takes the input of signals for voltage and current in power swing duration and estimates the phasors using phasor estimation techniques. After calculation of phasors, it carries out several operations, including relay characteristic design, fault phase differentiation, direction estimation, and fault detection. For the fast operation, some techniques are based on instantaneous values of voltage and current. In the next section, techniques for estimating direction and detecting faults are elaborated further.

9.4.1 Fault detection approaches during power swing

Detection of fault is the first task that must be performed by the relay so that further processes can be undertaken. Before performing the fault detection, voltage and current data measured by the relay should be sampled. Now, this sampled data is utilized to make the protection algorithm based on instantaneous values or to estimate the phasor using a phasor-based algorithm. There are many techniques for fault recognition within duration of a power swing. Some of these are discussed next.

9.4.1.1 Cumulative sum (CUSUM)-based technique [3]

CUSUM is a very powerful tool to perform fault detection. The CUSUM of negative sequence component-based technique is discussed which works in a series compensated environment also. To implement the technique, negative sequence current (I_2) can be obtained using three-phase current phasor measured at relay location and written as:

$$I_2 = \frac{I_a + \alpha^2 I_b + \alpha I_c}{3}, \tag{9.8}$$

where phasors I_a, I_b, and I_c are the phase current and $\alpha = e^{j2\pi/3}$. First, a derived signal Sn is obtained using negative sequence current and is expressed as:

$$S_n = \Delta \left| I_2(n) \right| = \left| I_2(n) \right| - \left| I_2(n-1) \right|. \tag{9.9}$$

After applying CUSUM on S_n, index g_n is calculated, as provided in eqn (9.10).

$$g_n = \max \left(g_{n-1} + S_n - \varepsilon, 0 \right). \tag{9.10}$$

Here, g_{n-1} is a delayed signal of g_n which is initialised by 0. ε is the drift parameter which is decided to keep g_n zero or negative during nonfault conditions. A fault is registered if

$$g_n > h. \tag{9.11}$$

Here, h is a constant and should ideally be zero.

9.4.1.2 Teager–Kaiser energy operator (TKEO)-based technique [2]

Previous technique is a current-based technique which is unaffected by capacitor coupling voltage transformer (CCVT) transients due to independence of voltage. However, current-based fault detection techniques are affected by current transformer (CT) saturation because current is zero if the CT core is entirely saturated. To cover all the protection aspects, an integrated technique is discussed which is based on TKEOs of average 3-phase instantaneous voltages and negative sequence current. TKEO of the average of 3-phase instantaneous voltage-based function (function-1) provided in eqn (9.14) works only for unsymmetrical ground faults, and the saturation of the current transformer has no effect on it. TKEO of negative sequence current-based function (function-2) successfully works for unbalanced and symmetrical faults and is not affected by CCVT transients.

For a discrete signal, the general expression for TKEO of any signal *x(n)* is written in eqn (9.12).

$$\Psi\left[x(n)\right] = x^2(n) - x(n-1)x(n+1). \tag{9.12}$$

To detect the fault, an average of three-phase instantaneous voltages (v_{avg}) is presented in eqn (9.13) and calculation of negative sequence current is already provided in eqn (9.8). Three-phase voltage averages can be expressed as

$$v_{\text{avg}} = \frac{v_a + v_b + v_c}{3}, \tag{9.13}$$

where v_a, v_b, and v_c are the instantaneous voltages measured at the relay location.

TKEOs of v_{avg} and I_2 (as function-1 and function-2) are provided in eqn (9.14) and (9.15).

$$\Psi\left[v_{\text{avg}}\right] = {v_{\text{avg}}}^2(n) - v_{\text{avg}}(n-1)v_{\text{avg}}(n+1), \tag{9.14}$$

$$\Psi\left[I_2\right] = {I_2}^2(n) - I_2(n-1)I_2(n+1). \tag{9.15}$$

Unsymmetrical ground faults are reported using TKEO of v_{avg} if

$$|\Psi[v_{\text{avg}}]| > 0. \tag{9.16}$$

The information about fault detection is provided by the magnitude of TKEO of I_2, and the necessary conditions for fault detection is specified as

$$|\Psi[I_2(n)]| > 0. \tag{9.17}$$

Only grounded unsymmetrical faults are recognized by function-2 in this approach; other fault types are not detected by TKEO of v_{avg}. Function-1 has high priority since it responds quickly. Outputs of function-1 and function-2 relate to an OR gate. It means that the integrated scheme provides the output if any one function works. The faults are identified by combining the output from the two techniques.

9.4.1.3 Park's transformation-based technique [5]

A Park's transformation-based technique for symmetrical fault detection during power swing is proposed by authors in [5]. By using Park's transformation theory, two signals with direct and quadrature (*d-q*) components are created from the three-phase voltage and current signals. Power coefficients are derived from the voltage and current components of d-q signals using a sample-to-sample moving window comparison.

Three-phase voltage signals are split into two components: direct and quadrature, as shown by

$$\begin{bmatrix} V_d(n) \\ V_q(n) \end{bmatrix} = \frac{2}{3}\begin{bmatrix} \text{Cos}\,(\theta) & \text{Cos}\,\left(\theta - \frac{2\pi}{3}\right) & \text{Cos}\,\left(\theta + \frac{2\pi}{3}\right) \\ -\text{Sin}\,(\theta) & -\text{Sin}\,\left(\theta - \frac{2\pi}{3}\right) & -\text{Sin}\,\left(\theta + \frac{2\pi}{3}\right) \end{bmatrix}\begin{bmatrix} V_a(n) \\ V_b(n) \\ V_c(n) \end{bmatrix}. \tag{9.18}$$

Here, the voltage signals in three phases are denoted by V_a, V_b, and V_c, whereas the components of the direct and quadrature axis are represented by V_d and V_q. Current signals in three phases, I_a, I_b, and I_c, can also be converted in the same manner. In the above equation, n represents the sample number and $\theta = n\omega\Delta t + \delta$, ω be representation of the angular frequency, Δt be the sampling period, and δ is denoted by the angle of V_d.

After determination of the voltage signals' direct and quadrature components, computation of the "C" coefficients of direct and quadrature components of voltage are provided in eqn (9.19) and (9.20).

$$C_{vd}(n) = V_d(n) - V_d(n-1), \tag{9.19}$$

$$C_{vq}(n) = V_q(n) - V_q(n-1). \tag{9.20}$$

Similarly, "C" coefficients from current signals are also obtained as

$$C_{id}(n) = I_d(n) - I_d(n-1), \tag{9.21}$$

$$C_{iq}(n) = I_q(n) - I_q(n-1). \tag{9.22}$$

Now, by multiplying these "C" coefficients of voltage and current components of the direct and quadrature axes, the power coefficients are derived.

Direct axis component is

$$C_{pd}(n) = C_{vd}(n) \times C_{id}(n). \tag{9.23}$$

And, quadrature axis component is

$$C_{pq}(n) = C_{vq}(n) \times C_{iq}(n). \tag{9.24}$$

During power swing, C_{pd} and C_{pq} values are nearly zero, but they become substantial when there are symmetrical fault situations. If C_{pd} and C_{pq} exceed the pre-established threshold, a fault is identified during swing, according to the detection criterion.

9.4.2 Fault direction estimation techniques during power swing

Determination of direction of fault is the next step after fault detection in protecting the power system. Limited research work has been done to determine the direction of a fault during a power swing. In this section, superimposed negative sequence components based-scheme and superimposed positive sequence components-based scheme are discussed for estimating the direction of a fault amid a power swing.

Figure 9.7(a,b) represents the superimposed negative sequence networks for forward (f_y) and backward (f_x) faults. Similarly, Figure 9.7(c,d) represents positive sequence networks for forward (f_y) and backward (f_x) faults.

From Figure 9.7(a,b) the relation of superimposed negative sequence voltage (SNSV) and superimposed negative sequence current (SNSC) for reverse and forward faults is provided using eqn (9.25) and (9.26).

$$\begin{aligned}\frac{\Delta V_{\mathrm{fx2}}}{\Delta I_{\mathrm{fx2}}} &= -(-Z_{\mathrm{C}}+Z_{\mathrm{BC2}}+Z_{\mathrm{SC2}}) \\ &= -(-Z_{\mathrm{C}}+Z_{\mathrm{BC1}}+Z_{\mathrm{SC1}}) = -Z_{\mathrm{fx2}}\end{aligned}, \tag{9.25}$$

$$\frac{\Delta V_{\mathrm{fy2}}}{\Delta I_{\mathrm{fy2}}} = (Z_{\mathrm{SA2}}+Z_{\mathrm{AB2}}) = Z_{\mathrm{fy2}}. \tag{9.26}$$

From Figure 9.7(c,d), the relation of superimposed positive sequence voltage (SPSV) and superimposed positive sequence current (SPSC) for reverse and forward faults is obtained using eqn (9.27) and (9.28).

$$\frac{\Delta V_{fx1}}{\Delta I_{fx1}} = -(-Z_C+Z_{BC1}+Z_{SC1}) = -Z_{fx1}, \tag{9.27}$$

$$\frac{\Delta V_{fy1}}{\Delta I_{fy1}} = (Z_{SA1}+Z_{AB1}) \;= Z_{fy1}. \tag{9.28}$$

It is proved mathematically from eqn (9.25) and (9.26) that the ratio of SNSV and SNSC is negative in backward direction and positive in forward direction. A similar observation is true for superimposed positive sequence components. Hence, angle of ratio of SNSV and SNSC provides the information of fault direction. Similarly, angle of ratio of SPSV and SPSC indicates the information of fault direction.

Figure 9.7 depicts the three-bus system with bus-A, bus-B, and bus-C. Line-BC is series compensated with reactance Zc. Relay is placed at bus B to protect the line-BC. The prefault voltage U_{AFD} is the voltage at fault and R_F be the fault resistance. The network as shown in Figure 9.7 is designed for

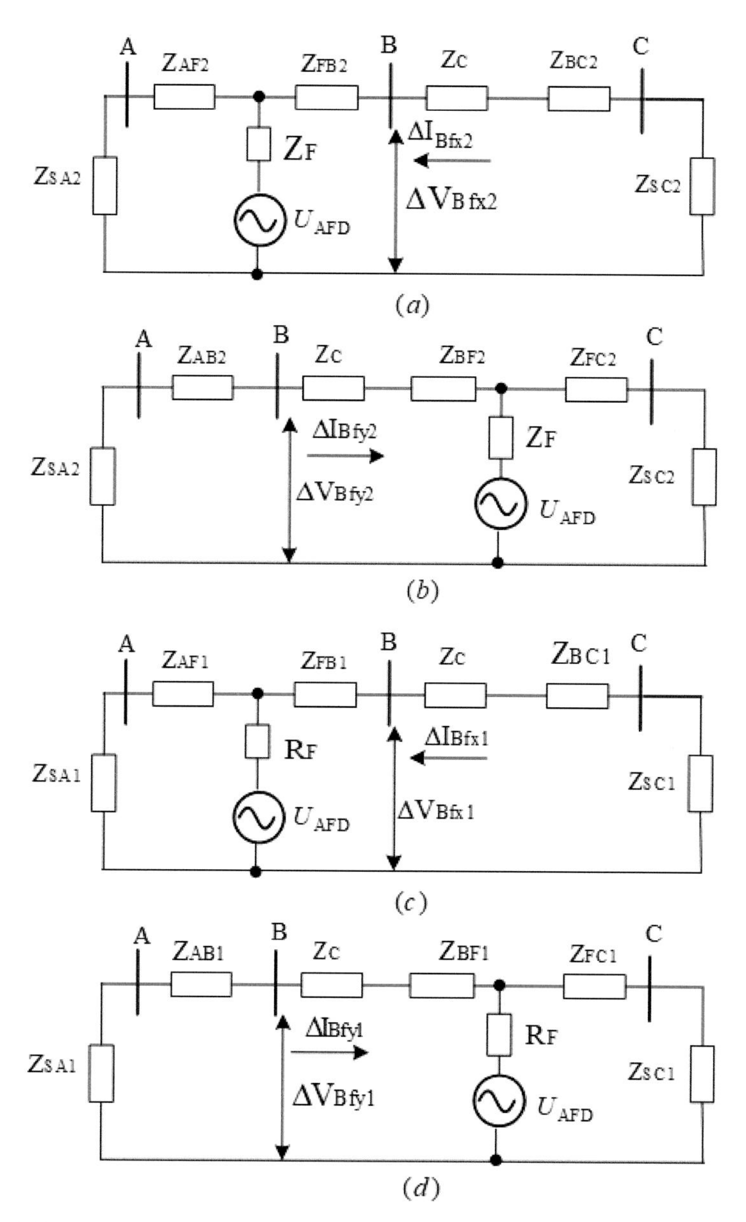

Figure 9.7 Networks of negative and positive sequence for the faults (a) reverse, (b) forward, (c) reverse, and (d) forward.

an ag-type (phase-a to ground) of fault. It is possible to derive the SNSV and SNSC phasors ΔV_2 and ΔI_2 as

$$\Delta V_2 = V_{Bfault} - V_{Bpre}, \tag{9.29}$$

$$\Delta I_2 = I_{Bfault} - I_{Bpre}. \tag{9.30}$$

Here the voltage and current pre-fault values depict V_{Bpre} and I_{Bpre}, while the voltage and current post-fault (either in the f_x or f_y direction) are V_{Bfault} and I_{Bfault}. The angle of ratio of SNSV and SPSC is defined as

$$\Delta\phi_2 = \angle\Delta V_{f2} - \angle\Delta I_{f2}. \tag{9.31}$$

The conditions for fault direction estimation are

$$\Delta\phi_2 > 0, \quad \text{backward fault}, \tag{9.32}$$

$$\Delta\phi_2 < 0, \quad \text{forward fault}. \tag{9.33}$$

In similar analysis, the SPSV and SPSC can be defined as

$$\Delta V_{f1} = V_{f1} - V_{pre1}, \tag{9.34}$$

$$\Delta I_{f1} = I_{f1} - I_{pre1}. \tag{9.35}$$

The expression for the phase difference between superimposed positive sequence voltage and current is as follows:

$$\Delta\phi_1 = \angle\Delta V_{f1} - \angle\Delta I_{f1}. \tag{9.36}$$

Hence, the conditions for fault direction estimation are

$$\Delta\phi_1 > 0, \quad \text{backward fault}, \tag{9.37}$$

$$\Delta\phi_1 < 0, \quad \text{forward fault}. \tag{9.38}$$

From eqn (9.32)–(9.33) and eqn (9.37)–(9.38), it is observed that the phase angles between superimposed positive and negative sequence voltages and currents are able to find out the direction of fault. During a symmetrical fault, negative sequence-based technique does not work due to unavailability of negative sequence components. However, a positive sequence components-based scheme works for symmetrical as well as unsymmetrical faults. A combination of both methods can also be implemented for detecting all types of fault cases.

9.5 Fault Section Identification during Power Swing for T-connected Transmission System [8]

In this section, fault section identification during power swing is discussed for T-connected transmission lines. Fault section identification issue is the combination of fault and its direction identification. For a normal transmission system, fault and its direction identification techniques can be combined together. However, a solution of fault section identification for T-connected transmission networks has been provided in this section.

Here, a communication-based technique is discussed for three terminal lines [8], which identify the fault portion of a three-terminal system in power swing duration. Communication system provides the data of all buses to the processor which synchronizes with the GPS system. After processing the data of all buses, the protection algorithm is executed and the relay will generate the trip signal [8]. Here, this technique is based on SPSCs and makes use of a communication-aided protective relay. It detects faults by using synchronized current and voltage phasors from all three buses M, D, and P, as illustrated in Figure 9.8.

In this method, the peak magnitude of the SPSCs calculated at T-point and the SPSCs observed at three terminals are compared. The estimated currents at T-point are measured using a long line model utilizing the ABCD parameter

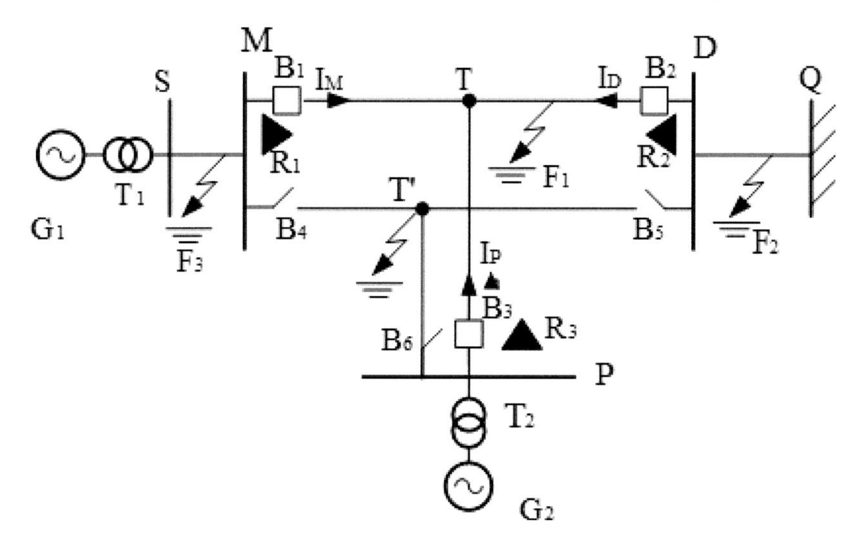

Figure 9.8 A 400 kV, 50 Hz double-circuit three-bus system.

approach. The parameters I_{M1}, I_{D1}, I_{P1} are positive sequence currents and positive sequence voltages are identified as V_{M1}, V_{D1}, and V_{P1}, respectively, at buses *M, D,* and *P*. Using synchronized voltage and current data captured at buses M, D, and P, $I_{TM1}, I_{TD1}, I_{TP1}$ represent estimated values of positive sequence currents at T-point. As per the extended line ABCD parameter technique, the values of I_{TM1}, I_{TD1}, and I_{TP1} are calculated as

$$I_{TM1} = -\frac{V_{M1}}{Z_C} Sinh(\gamma l_{MT}) + I_{M1} Cosh(\gamma l_{MT}), \tag{9.39}$$

$$I_{TD1} = -\frac{V_{D1}}{Z_C} Sinh(\gamma l_{DT}) + I_{D1} Cosh(\gamma l_{DT}), \tag{9.40}$$

$$I_{TP1} = -\frac{V_{P1}}{Z_C} Sinh(\gamma l_{PT}) + I_{P1} Cosh(\gamma l_{PT}). \tag{9.41}$$

Here γ is the line's propagation constant per unit length, and Z_C is the line's characteristic impedance. The distances of buses M, D, and P from T-point are given by the parameters l_{MT}, l_{DT}, and l_{PT}, respectively.

The measurements of the SPSCs at the relays at buses M, D, and P are obtained as

$$I_{M1\_S}(n) = I_{M1}(n) - I_{M1}(n - N), \tag{9.42}$$

$$I_{D1\_S}(n) = I_{D1}(n) - I_{D1}(n - N), \tag{9.43}$$

$$I_{P1\_S}(n) = I_{P1}(n) - I_{P1}(n - N), \tag{9.44}$$

where $I_{M1\_S}(n)$, $I_{D1\_S}(n)$, and $I_{P1\_S}(n)$ are the SPSCs at buses M, D, and P as of this moment *n*. The sample count in the unit cycle is indicated by the letter *N*.

The highest value of these three SPSCs is established by

$$h_1(n) = \max\left(\left|I_{M1\_S}(n)\right|, \left|I_{D1\_S}(n)\right|, \left|I_{P1\_S}(n)\right|\right). \tag{9.45}$$

In a similar manner, the T-point SPSCs estimates are computed as

$$I_{TM1\_S}(n) = I_{TM1}(n) - I_{TM1}(n - N), \tag{9.46}$$

$$I_{TD1\_S}(n) = I_{TD1}(n) - I_{TD1}(n - N), \tag{9.47}$$

$$I_{TP1\_S}(n) = I_{TP1}(n) - I_{TP1}(n - N), \tag{9.48}$$

where $I_{TM1\_S}(n)$, $I_{TD1\_S}(n)$, and $I_{TP1\_S}(n)$ are the estimated SPSCs at T-point based on the voltage and current phasors of terminals M, D, and P, in that order.

These three SPSCs' peak values are established by

$$h_2(n) = \max(|I_{TM1\_S}(n)|, |I_{TD1\_S}(n)|, |I_{TP1\_S}(n)|). \tag{9.49}$$

For an internal fault F_1 in line DT, the voltage and current distribution fault components cause the current measured at terminal D to exceed the currents recorded at terminals M and P. It is evident from Figure 9.9(a) that $h_1(n)$ is observed similar as $|I_{D1\_S}(n)|$ and $h_2(n)$ is observed similar as $|I_{TD1\_S}(n)|$.

It is also observed that

$$|I_{TD1\_S}(n)| > |I_{D1\_S}(n)|, \tag{9.50}$$

$$h_2(n) > h_1(n). \tag{9.51}$$

Therefore, eqn (9.51) relies on the internal fault. The fault is situated between T-point and terminal D, according to the maximum magnitude of

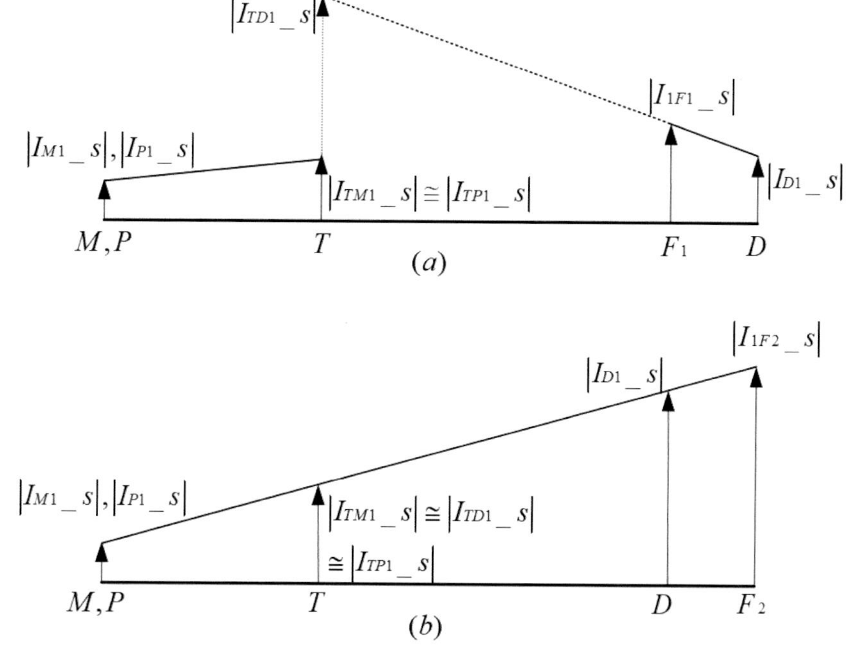

Figure 9.9 Superimposed currents calculated at T-point and observed at buses; (a) internal, (b) external faults.

predicted SPSCs at T-point (calculated using terminal D's voltage and current data). This also applies to issues near other endpoints.

Now, for the case of an external fault near terminal D, the value of current measured at terminal D is increased but the currents in lines MT, DT, and PT remain unaffected. From Figure 9.10(b), the value of $h_1(n) = |I_{D1\_S}(n)|$ and $h_2(n) = |I_{TM1\_S}(n)| \approx |I_{TD1\_S}(n)| \approx |I_{TP1\_S}(n)|$.

Consequently, in the case of an external fault, it is noted that

$$h_2(n) < h_1(n). \tag{9.52}$$

The deviation between indices h_2 and h_1 is calculated to establish a new index Φ_2 as

$$\Phi_2(n) = h_2(n) - h_1(n). \tag{9.53}$$

Owing to signal modulation during power swing, Φ_2 has a very modest nonzero value and index h_2 grows slightly from index h_1. This small nonzero value is considered as a threshold, which is indicated by η_1. η_1 indicates that this little nonzero number is regarded as a threshold. Thus, the criteria specified to identify internal fault is as

$$\Phi_2(n) > \eta_1. \tag{9.54}$$

After indication of internal fault, there are some checks to identify the faulted section. The conditions for fault section identification are

$$\text{If } h_2(n) = \begin{cases} I_{TM1\_S}(n) & \text{Line MT is a fault section} \\ I_{TD1\_S}(n) & \text{Line DT is a fault section} \\ I_{TP1\_S}(n) & \text{Line PT is a fault section} \end{cases}. \tag{9.55}$$

Using eqn (9.55), the faulted section of the power system is identified.

9.6 Conclusion

In this chapter, protection challenges during power swing and its solutions are discussed. Mathematical analysis of reduction of intensity of power swing by series compensation is provided. Various advanced techniques of fault detection such as CUSUM, TKEO-based techniques are discussed which are very effective during power swing. Fault direction estimation techniques based on superimposed negative sequence and superimposed positive sequence components are discussed and its mathematical analysis is also provided using local parameters. Fault section identification issue for T-connected transmission system is resolved by combining fault detection technique and fault direction estimation technique. Overall advanced protection issues with its mathematical analysis during power are discussed in this chapter.

References

[1] Power Syst. Relay Committee Report, Power swing and out of step considerations on transmission lines, IEEE PSRC WG D6, pp.1-59, 2005.
[2] J. Kumar, P. Jena, "Solution to Fault Detection During Power Swing Using Teager–Kaiser Energy Operator," Arabian Journal for Science and Engineering, vol. 42, no. 12, pp. 5003-5013, Apr. 2017.
[3] H. Shekhar, J. Kumar, A. Nayak, "A Fault Detection Technique during Power Swing in a TCSC-Compensated Line Using Teager Kaiser Energy Operator," IETE Journal of Research, pp. 1-21, 2020.
[4] P. K. Nayak, A. K. Pradhan, P. Bajpai, "A Fault Detection Technique for the Series-Compensated Line During Power Swing," IEEE Transactions on Power Delivery, vol. 28, no. 2, pp. 714-722, Apr. 2013.
[5] K. Andanapalli, B. R. K. Varma, "Park's transformation based symmetrical fault detection during power swing," Eighteenth National Power Systems Conference (NPSC), Guwahati, pp. 1-5, 2014.
[6] J. Kumar, P. Jena, "Fault direction estimation during power swing," IEEE 1st International Conference on Power Electronics, Intelligent Control and Energy Systems (ICPEICES), Delhi, pp. 1-3, 2016.
[7] H. Shekhar, J. Kumar, "Recognition of Fault Section during Power Swing in the Presence of TCSC Environment," Arabian Journal for Science and Engineering, 2022.
[8] H. Shekhar, J. Kumar, "Fault Section Identification of a Three Terminal Line during Power Swing." International Transactions on Electrical Energy Systems, vol. 31, no. 4, pp. 1-25, 2021.

10

Conclusion

V.K. Sood[1], O.H. Gupta[2]

[1]Ontario Tech University, Canada
[2]Department of Electrical Engineering, NIT Jamshedpur, India
Email:vijay.sood@ontariotechu.ca, omhari.ee@nitjsr.ac.in

Abstract

The book is composed of ten carefully selected chapters by a variety of authors to highlight the impact of soft computing methods in power systems. A summary of these chapters' contents is presented.

Keywords: Power electronics, soft computing, power generation, power transmission and distribution

10.1 Introduction

This book covers applications of soft computing applications for advancements in power systems. Over the past 120 years or so, the power system has grown from a simple radial system with a few generators and dedicated loads to a complex and integrated power system that mankind cannot live without. The rapid growth in power systems took place after the Second World War in 1945, as long-distance AC lines were required from remote generating sites with ever increasing higher voltages to reduce the transmission power losses. The early 1950s saw the introduction of HVDC transmission systems in Sweden using mercury arc converters for the special needs of bulk power transmission over long distances and undersea interconnections. The introduction of power electronics in the late 1960s brought about the era of static var compensation and flexible AC transmission systems (FACTS).

The widespread merging, usage and cost reduction of power electronics, information technology and communications facilities in the 1980s brought about another revolutionary change to the industry as the control and protection of the ever-growing power systems became increasingly important and complicated [1].

The era of climate change, environmental concerns, and the gradual replacement of fossil fuels as the source of electrical energy has now reached the point where the smart grid is now becoming highly necessary to any national grid. The modern power system is now intimately involved with power electronics, HVDC systems, microgrids, power system protection, and energy scheduling to reduce the cost of electricity and provide a stable controllable power supply.

The book is composed of ten carefully selected and diverse chapters by a variety of authors to highlight the impact of soft computing methods in power systems. A summary of these chapters' contents is presented next.

10.1.1 Chapter 1

Chapter 1 covers how soft computing methods have grown over the years, influencing all pertinent fields of science and engineering. The soft computing methods, inspired by the human mind and biological behaviour, have proven to be excellent tools to overcome the difficulties faced in a vast variety of applications in the area in existing power systems. In recent times, power electronics-based converters with distributed generations and loads are being deeply inducted into the traditional power system. With the inclusion of these modern power system components, the applications of soft computing in the various fields of power systems will have to undergo major modifications or upgrades in providing the solutions for the modern power system problems. This chapter provides the information regarding the power system areas where power electronics is making a significant impact.

10.1.2 Chapter 2

Chapter 2 presents different soft computing techniques such as fuzzy logic, neural networks, support vector machine, and nontraditional optimization techniques. The chapter provides details of these techniques, including their working principles.

10.1.3 Chapter 3

Chapter 3 discusses a load flow (LF) methodology for an AC-DC radial distribution system. A per unit model of a power system converter is introduced to bridge the difference between an AC and DC distribution systems to carry out the LF of the distribution system. The different ways that power converters can operate has also been considered to show the effectiveness of the indicated approach. The sensitivity and breakpoint matrices have been designed to access the LF solution of DS with PV and Vdc bus. The indicated method is computationally effective due to the concise algebraic operation and effective search technique.

10.1.4 Chapter 4

A power flow study is an operation for the numerical analysis of the power flow in an interconnected power system. A power-flow study uses simplified notations such as a single-line diagram and per-unit system, and focuses on various aspects of AC power parameters, such as voltages, voltage angles, real and reactive power. Since the power flow study is a steady-state operation, a three-phase solution is not needed if a balanced system is considered. Power flow studies are vital for the planning of future expansion of power systems as well as in determining the best operation of existing systems. This chapter focuses on sample studies using an IEEE 14-bus test system using Matlab environment.

10.1.5 Chapter 5

Chapter 5 proposes a novel hybrid optimization technique for day-ahead scheduling in a smart grid. A grid-connected or islanded microgrid made up of distributed energy sources (DERs) requires a power management/dispatch system to control the power dispatch and meet the load demand in the system. At the tertiary control level in a typical microgrid, an optimal scheduling mechanism is used to manage the power generated from the local DERs, energy drawn from the grid and energy consumption by the load. A hybrid feedback PSO-MCS algorithm is implemented using swarm intelligence and cuckoo search to enhance the performance and obtain a cost-effective solution for a microgrid prosumer. A comparison has been made of the hybrid feedback PSO-MCS (HFPSOMCS) algorithm with PSO and modified CS (MCS) algorithm. The best performing algorithm among the three is executed in MATLAB/Simulink and Python IDE platforms to compare the execution time.

10.1.6 Chapter 6

Chapter 6 covers how photovoltaic (PV) systems have grown in importance over the last decade. Due to their unique designs and growing commercial interest, grid-connected single-phase transformerless (TL) solar inverters are now being explored further. There are several benefits over transformer galvanic isolation-based inverters, including cheaper prices, lower weight, smaller volume, higher efficiency, and less complexity. In this chapter, transformer less "grid-connected" inverters with negative, positive, and zero cycle operations are explored. A detailed examination of various topologies is carried out. Simulated results are shown to verify the effectiveness of the systems.

10.1.7 Chapter 7

Chapter 7 covers the influence of HVDC systems in power transmission networks worldwide. Due to the fast, precise, and flexible control of HVDC links for controlling DC link current, HVDC systems play a significant role in modern power systems and smart grids. The HVDC systems are classified into different types based on application, modelling, and control features. The LCC-HVDC system is basically used for bulk power transmission and VSC-HVDC is preferred for integration of renewable energy sources (RES) in a smart grid. Thus, to achieve successful operation of HVDC links with RESs, extensive planning and research should be performed before installing HVDC to integrate renewable sources. An incorrect sizing of the HVDC converter may lead to unnecessary curtailment of renewables or underutilization of HVDC converters. The provisions for a virtual generator and load shedding options are highly recommended in a power system with extensive RES penetration. This is due to low inertia of modern smart grids and RESs hardly respond to any frequency instability due to HVDC links. Thus, together with the promising features of HVDC system there are several other issues that also arise due to evolving changes ongoing modification in a power system. Some of these are issues related to optimization and cost-effectiveness in operation, electromagnetic interference, security, day-ahead operational strategies to maximize the accommodation of renewable energies and challenges related to fault detection and diagnosis. The low cost, traceable, and simple solutions to such issues can effectively be provided by using soft computing (SC) techniques. This chapter summarizes the results of a few SC methods applied in HVDC systems to resolve various operational, control, and protection problems. Finally, real-time simulation results, and hardware-in-loop (HIL) model of a VSC-HVDC test system is presented.

10.1.8 Chapter 8

Chapter 8 covers how, for any nation, the energy sector is a vital critical infrastructure which plays a significant role in its development and economy. A high-quality, reliable, and resilient electricity supply is a necessity in the modern world. To fulfil the requirement, the concept of the smart grid is introduced which is viable due to overwhelming advancement in power electronics converters, smart sensors/meters, and information and communication technology (ICT). This chapter discusses the principle of a smart grid cyber–physical system operation. It deals with the cybersecurity of smart grid infrastructure issues, selection of communication technologies in smart grid, cybersecurity standards, typical type of cyberattack and threats against smart grid, and impact of a cyberattack on smart grid. It explains how a cyberattack handling approach is different from information technology (IT) to operational technology (OT).

10.1.9 Chapter 9

Chapter 9 presents the power swing is the condition where three-phase power varies with power angle. This change in power angle can occur due to a sudden load change, line switching, and/or generator disconnection, etc. Variation in power angle causes additional frequency components to be introduced in the system. Due to this phenomenon, distance relay may malfunction within the duration of the power swing. To prevent such malfunctioning, the distance relay is blocked by the blocking signal generated by a power swing blocker (PSB). During this period, the transmission system remains unprotected which is undesirable. An additional relay can be utilized with suitably designed protection algorithms within the period of power swing. If the line is series compensated, then the protection challenge becomes more complex during the power swing. In this chapter, fault detection and fault direction estimation issues have been discussed. The combination of these two issues makes the fault section identification, which is also deliberated in this chapter.

10.2 Conclusion

Overall, this book is rich in discussing real applications of soft computing (SC). These SC methods, inspired by the human mind and biological behaviour, have proven to be excellent tools to overcome the difficulties faced in a variety of applications in power systems and related areas and these methods are robust and provide low-cost solutions. Nowadays, SC methods, considered as a branch of AI, have grown over the years in almost all fields of science and engineering. SC techniques offer an effective solution for studying and modelling the behaviour of renewable energy generation, operation of grid-connected renewable energy systems, and sustainable decision-making among alternatives. The tolerance of SC techniques to imprecision, uncertainty, partial truth, and approximation makes them useful alternatives to conventional techniques.

Furthermore, the rapid growth in SC technologies plays an important role in powerful representation, modelling paradigms, and optimization mechanisms for solving power system issues such as power quality, reactive power control, oscillation and stability problems, renewable energy resource evaluation, design of energy efficiency systems, economic load dispatch problems, or very different energy system applications in smart grids.

10.3 Acknowledgements

The editors are grateful to all chapter authors who have been willing and helpful to provide their valuable insights into these topics discussed here. The editors also thank staff members of River Publishers who provided their services and guidance in putting this book together in a timely and efficient manner.

References

[1] Jigneshkumar Patel and Vijay K. Sood, "Review of Digital Controllers in Power Converters", IEEE EPEC 2018, Toronto, Canada. Oct 2018.

Index

About the Editors

Vijay Sood received his Ph.D. degree from the University of Bradford, UK, in 1977. He is currently a Professor at Ontario Tech University, Canada. Previously, he was a senior researcher at the Research Institute of Hydro-Quebec (IREQ, Montreal). He has extensive experience in the simulation of HVDC-FACTS systems and their controllers. He has authored two textbooks on HVDC transmission. His research focuses on the monitoring, control, and protection of power systems and on the integration of renewable energy systems into smart grids. Dr. Sood is a Registered Professional Engineer in the province of Ontario, Canada. He is a Life Fellow of the IEEE, Fellow of the Engineering Institute of Canada, and Emeritus Fellow of the Canadian Academy of Engineering. He served previously as a Director of the IEEE Canadian Foundation and as a former Editor of the IEEE Transactions on Power Delivery, and Co-editor of the IEEE Canadian Journal of Electrical and Computer Engineering. He is currently the Editor-in-Chief of the Power series for River Publishers.

Krishna Murari received the B.Tech degree in electrical engineering from WBUT, India, in 2010, the M.E degree in power systems from Thapar University, India in 2014, and the Ph.D. degree in electrical engineering from the Indian Institute of Technology Roorkee, India in 2019. He currently serves as a Research Assistant Professor in the Department of Electrical Engineering and Computer Science at The University of Toledo, Toledo, Ohio, USA. Prior to this, he worked as a Research Associate at Clarkson University, Potsdam, New York. He has also worked as a Postdoctoral Fellow with the Energy Production, and Infrastructure Center (EPIC), University of North Carolina at Charlotte, North Carolina, USA. He is the recipient of the third-best prize award from The Industrial Automation and Control Committee of the IEEE Industry Application Society for his research work presented at the IAS annual meeting 2021. He is a reviewer of journals, such as IEEE, IET, MDPI, and Taylor Francis journal. His research interests include power system analysis, power flow studies, demand-side management, optimization and control of DERs, optimal power flow, distribution network pricing, and smart grid.

Om Hari Gupta received the M.Tech. degree in power electronics & ASIC design from Motilal Nehru National Institute of Technology, Allahabad, Prayagraj, Uttar Pradesh, India, in 2011 and the Ph.D. degree in electrical engineering from Indian Institute of Technology Roorkee, Roorkee, Uttarakhand, India, in 2017. He is an awardee of the Canadian Queen Elizabeth II Diamond Jubilee Scholarship for research visiting Canada and from March 2017 to June 2017, he was a visiting Researcher at the Ontario Tech University, Canada. Since 2018, Dr. Gupta is an Assistant Professor with the Department of Electrical Engineering, National Institute of Technology Jamshedpur, Jharkhand, India. He is one of the founding organizing secretaries of the conference series "Electric Power and Renewable Energy Conference - EPREC". Dr. Gupta is the author/editor of five books and has more than 80 publications including one Indian Patent (Granted). He is a senior member of IEEE and a reviewer of many reputed journals like IEEE Transactions on Power Delivery, IET Generation and Transmission, Electric Power Systems Research, Electric Power Components and Systems, and International Journal of Electrical Power and Energy Systems. His research interests include power system protection, microgrid, renewable-based distributed generation, and electric power quality.

Anupam Kumar currently serves as an Assistant Professor in the Department of Electronics and Communication Engineering (ECE), National Institute of Technology Patna, India, from Oct 2022. Prior, he was an Assistant Professor at IIIT Kota and IIIT Bhagalpur, India, from 2019–2022 and 2018–2019, respectively. His current research interests include Fuzzy logic systems, Robotics, Signal Processing, Robot Assistive device, Biomedical engineering, Health care application, Deep learning, Machine learning, etc. He received the M.Tech and Ph.D. degree from IIT Roorkee, India, in 2012 and 2018 in the ECE Department. He has published more than 14 SCI/Scopus journals, 13 conferences, 7 book chapter, 2 book chapters, 1 patent granted and 2 patent published. His current google scholar citation is more than 600. He is serving as reviewer in journals like IEEE Transactions, Elsevier, Springer, Taylor Francis journal and Wiley. He also served as a reviewer at many conferences. He is also a recipient of the best paper award at NIT Kurukshetra. He has attended several workshops and participated in many research activities.

Printed in the United States
by Baker & Taylor Publisher Services